D1825028

Contents

Editorial

Rachel Masika

The global climate is subject to increasing change, and this has become more evident over recent years.[1] In particular, the atmospheric concentrations of greenhouse gases have increased, augmenting global warming. These concentrations of carbon dioxide, methane, and nitrous oxide are higher now than at any time during the last thousand years, and the weight of scientific evidence suggests that observed changes in the earth's climate are at least in part due to human activities (IPCC 2001a).

The latest climate projection models of the United Nations Inter-governmental Panel on Climate Change (IPCC), a high-level, independent, scientific advisory body, suggest that if no action is taken now to reduce greenhouse gases, air surface temperatures could rise to levels that will significantly alter ecosystems. The IPCC and other forecasters predict that such global warming could result in the following changes:

- a rise in sea levels exposing many communities to severe flooding from storm surges;
- a decline in water availability with billions of people facing water shortages in the Middle East and the Indian subcontinent;
- disruptive seasonal rainfall patterns which will bring about droughts and floods, dramatically reducing crop yields and increasing food insecurity in much of the developing world;
- increased frequency and intensity of extreme weather events leading to loss of life, injury, mass population dislocations, and economic devastation in poor countries;
- a decline in human health as people's resistance to disease is weakened by heat stress, water shortages, and malnutrition. Increases in infectious diseases and waterborne illnesses, and higher levels of pollution leading to a rise in respiratory illnesses, will be widespread (IPCC 2001b; Martens 1998).

No one will be immune from the overall impacts of climate change, but it will have a disproportionate effect on the lives of poor people in developing countries, where poverty increases people's vulnerability to its harmful effects. Poor people in overcrowded temporary settlements erected on unsuitable land will be at risk of flooding and land slides. Those eking out an existence from subsistence farming will have no savings or assets to insure them against external shocks. Lack of sanitation and limited access to clean water, poor diet, and inadequate healthcare provision will undermine resistance to disease. A lack of social status and the remoteness of some settlements means that some people will not receive adequate warnings of impending disasters, and relief efforts will be less likely to reach them. Lack of education, and official neglect, means that poor and marginalised people will have little alternative but to remain in, or return to, disaster-prone areas, with diminished assets.

Adverse changes in climate are likely to exacerbate the problems that developing countries are already facing, particularly since 94 per cent of the world's major natural disasters between 1990 and 1998 occurred in developing countries (Oxfam 2000, 1). It is important to acknowledge, however, that some climate changes may have beneficial impacts. As Terry Cannon points out in his article in this collection, increased floods – a projected consequence of climate change – have the potential to improve rather than destroy livelihoods. Inundation and silt can improve soil fertility, thus boosting crop yields. Flooding can also create ponds, improving conditions for fish breeding. Cooler, temperate climates, particularly in the Northern hemisphere, might also benefit from warmer temperatures.

The opportunities and challenges that men and women will face in responding to climate change are uncertain and unclear, as are the precise shape, form, scope, time-scales, and impacts of projected climate variability in different environmental settings. Given this degree of uncertainty, many may be tempted to ask of gender analysts and planners, why should we worry about this? There are difficulties in assessing the gender-differentiated risks and outcomes of ecological changes, particularly when the debate takes place within a highly-charged political environment where the validity of scientific hypotheses is questioned, and the threats are not fully understood. However, an understanding of potential gender-differentiated impacts can be gleaned from studies of gender issues in environmental and disasters management, where these demonstrate how individuals and communities are affected by and respond to environmental problems in different ways.

Gender issues have not been considered in wider climate change discourses and initiatives. The international response to the implications of climate change has largely focused on mitigation initiatives (the reduction of greenhouse gases), and has directed less attention to adaptation strategies (assistance with adapting to the adverse impacts of climate change on food, livelihood, and human security). Predominant approaches and policy responses have focused on scientific and technological measures to tackle climate change problems. They have displayed scant regard for the social implications of climate change outcomes and the threats these pose for poor men and women, or for the ways in which people's political and economic environments influence their ability to respond to the challenges of climate change. While scientific approaches remain crucial, this collection of articles argues that political and socio-economic issues must be taken into consideration, particularly since the climate change agenda is set by the rich and powerful, and can exclude the interests of the poor and less powerful within their variously constrained contexts.

The authors in this collection explore the connections between gender, poverty, and climate variability, and debate whether, why, and how gender and poverty issues matter within the 'climate change policy process'. Many discuss the vulnerability to the harmful risks of climate change that gender and poverty bestow, and demonstrate why these areas need to be considered and integrated into climate change interventions (policy debates, international agreements, and national and local programmes). Some explore why these issues have been largely neglected in research and analysis to date, suggest how they can be most effectively tackled, and which approaches can aid this process.

Gender or poverty?

It is widely accepted that the distribution of vulnerability to climate hazards and environmental degradation is not equal across societies and countries. Although location-specific climatic patterns are key factors in assessing risks and threats, levels of infrastructure, economic development, social equality, and political influence of countries and communities will affect the extent of their vulnerability to climate fluctuations. Adaptation – the ability of human systems to adapt to and cope with change – depends on factors such as wealth, technology, education, information, skills, infrastructure, access to resources, and management capabilities (IPCC 2001, 8). The adaptive capacity of men and women to environmental degradation will depend largely on the extent to which they can draw on these variables within varied contexts. Those with the least resources have the least capacity to adapt and are the most vulnerable.

The relative importance of gender or poverty in understanding vulnerability to climate change threats is subject to debate. A point of conjecture in this collection is the question of how far gender or poverty influences men and women's vulnerability to ecological risks and their capacity to respond to climate variation in localised contexts, and the extent to which poverty or gender should be the entry point for vulnerability reduction measures. In his article on the gender dimensions of climate hazards in Bangladesh, Terry Cannon asserts that the evidence for whether the impacts of hazards are worse for women is inconclusive and variable. Margaret Skutsch, also in this collection, suggests that poverty is the main variable, and that the issue of women's vulnerability to climate risks is best tackled through gender-responsive poverty reduction measures.

Other contributions to this collection highlight the central role that gender inequalities play in women's sensitivity to climate shocks, and their capacity to respond. Fatma Denton explains why gender matters to climate change processes, while Irene Dankelman's review demonstrates the significant role that gender relations – through their role in influencing which resources women or men can access – play in determining sensitivity to climate change and environmental degradation, and their capacity to cope with the outcomes.

There is still a tendency to link women with poverty, and by inference, to vulnerability. This conflation of poverty with women masks the underlying conceptual and structural underpinnings of gender inequality and poverty which, although closely associated, are not synonymous. Roy and Venema illustrate how close – and sometimes inextricable – these associations are, in their article presenting the 'capabilities approach' as a way of reducing poverty and gender inequalities in people's ability to adapt to climate change pressures.

Gender-specific implications of climate change outcomes

Some authors explore the gender-specific implications of climate change outcomes on human, food, and livelihood security, drawing on gender analyses of natural disasters, environmental conservation, and rural livelihood systems. The implications of gender divisions in labour for coping with environmental change, and the different ways in which men and women respond to disasters, are examined. These differences, largely due to unequal social relations, determine men's and women's roles, behaviour, and responsibilities in the household, workplace, and community. They determine their access to income to tap into material and productive resources that can provide security, protection, and recovery; and they determine individuals' power to influence or control events and outcomes that affect them. These differentiated power relations and levels of access to resources are key to understanding men's and women's vulnerability, their exposure to risk, their coping capacity, and their ability to recover.

More specifically, exposure to risk is influenced by existing nutritional inequalities, restricted mobility, and practices associated with understandings of 'appropriate' behaviour for men and women. Cannon highlights the case of Bangladesh, where women's lower nutritional status in pre-disaster situations worsened during crises. Nelson *et al.* cite the example of how a disproportionate number of women died in the 1991 cyclone in Bangladesh because of cultural norms concerning the preservation of female honour that meant they left it too late to leave their homes, and were less likely than men to know how to swim.

The capacity to cope and the potential to recover from climate shocks are affected by access to material and productive resources such as income and employment. Nelson *et al.* point out how in the aftermath of Hurricane Mitch in Honduras and Nicaragua, it was more difficult for women to return to work because of increased domestic and care responsibilities. Where drought and desertification, considered to be slow-onset disasters, make male migration in search of employment necessary, women's domestic and care responsibilities can increase, making it difficult for them to engage in income-generating work. Nonetheless, the widely-held view that male out-migration makes women more vulnerable is questionable because in some instances male migration can give women greater decision-making powers, and open up new livelihood possibilities for them.

Considering gender divisions of labour in agriculture, fisheries, the informal sector, the household, and the community, can assist us in pinpointing where vulnerability to ecological threats lies. Women's dependence on communal tree and plant resources, and their responsibility for fetching water, can place them under increasing strain as they trek further in search of firewood, and face diminishing plant resources and water shortages. Trekking long distances for water and fuel also affects the academic performance of young girls. Girls are often kept at home to help with household duties, and this is particularly the case in times of household stress or high workload. Similarly, a disproportionately large number of women work in the informal sector, and informal sector jobs are often the worst hit and slowest to recover when disasters strike.

Climate changes have the potential to create widespread additional health problems. These are likely to increase women's workloads still further, since women have traditionally had responsibility for caring for the sick and the elderly. Women's health may also suffer as a result of their existing lower access to health services, reduced nutritional status, and the requirement on them to juggle multiple roles. Men's mental health may also suffer, as men are less likely to seek counselling for trauma, a possible outcome of experiencing disaster. Rosa Reyes highlights the case of the 1997-8 El Niño events in Peru, where malnutrition among women was a major cause of peri-partum illness.

Current responses: mitigation and adaptation

International responses to climate change risks have tended to reflect the priorities and interests of wealthier countries, with the majority of the responses focused on greenhouse gas mitigation at country level, at the expense of adaptation measures and support. Mitigation has revolved around the reduction of greenhouse gases (through large-scale technical initiatives) and the enhancement of natural carbon sinks – particularly forests – to absorb them. However, attempts to develop mitigation strategies expose the power inequities within the international system. The unilateral position taken by the USA, a major greenhouse gas emitter, in refusing to endorse the Kyoto agreement, demonstrates how the 'big' business and economic considerations of powerful countries can derail mitigation efforts, and the importance of political and economic power in the success of mitigation efforts.

The significance of equity issues (social status, wealth, and power) in the resilience of human systems and security (food, livelihoods, health, and survival) to environmental change has often been over-looked in climate change interventions. A focus on technical solutions to climate change problems has ignored the social and political aspects of finding appropriate solutions. Many of the concerns of poor countries fall within the 'adaptation' scenario, in part due to their dependence on the physical environment for their food and livelihood security, and their limits in providing adequate protection against the shocks of climate disasters such as the recent floods in Mozambique and Bangladesh.

Adaptation is a key focus of this collection, and is also the area where gendered differences are most stark. Assisting those most vulnerable to climate risks requires an understanding of the complex and often intertwined influences that power, wealth, and social status have on who is most affected, who has the capacity to cope, and who decides on what action is to be taken. Ignoring these variables in environmental interventions risks exacerbating human suffering and reinforcing inequality between men and women and across countries. Including these considerations in the formulation and implementation of adaptation and mitigation measures can enhance adaptive strategies and assist in promoting gender equality.

Environmental NGOs and other pro-poor climate change lobbyists have played a central role in drawing policy-makers'

attention to equity and poverty issues. While gender advocates have played a key role in highlighting gender disadvantage and its influence on the success or failure of environmental and relief interventions, assumptions about men and women continue to translate into initiatives that place greater burdens on women's time and labour without rewards, and do not provide them with the inputs (education, information, and land rights) they require. Where gender issues have been considered, their integration into programmes has been insufficient in addressing the transformative requirements of social change that are inherent in addressing gender inequality and giving greater visibility to women's contributions to environmental conservation.

Natural resource management pro-grammes tend to rely too heavily on women's conservation capacities because it is assumed that they are naturally predisposed to serve their community by protecting the environment on which they depend for their livelihoods. As Dankelman's article demonstrates, rural women in developing countries are disproportionately adversely affected by environmental degradation. Nelson *et al.*, however, caution against assumptions that rural women are 'closer to nature', since such ideas can entail that development projects rely too heavily on women's unpaid labour. In addition, Denton cautions against using women as promotional agents for environmental conservation and tree-planting schemes without addressing other aspects that are important to good resource management and environmental conservation, such as women's ability to own land.

Initiatives that fail to address the transformations required for social change, or to challenge gender discrimination and disadvantage, reinforce gender inequality and miss the opportunity to utilise and enhance poor rural women's and men's skills and indigenous knowledge effectively. Emily Boyd demonstrates how a climate mitigation project in Bolivia contributed to gendered differences in outcomes, and traces the patriarchal relationships, from global decision-making processes to local impacts, that contributed to this. In this particular case, although women's practical needs were at least partially met by the project, their strategic interests were overlooked. Reyes shows that whereas women take responsibility for community organising at local levels in Peru, their under-representation in wider political rights-based movements, and at official national and regional levels, means that the impacts of their experience, skills, and environmental knowledge can be limited.

Strategies for gender-responsive policy and practice

Because the major causes of human-related climate variability are linked to energy use, greenhouse gas mitigation has been an essential part of international strategies to offset the risks of climate change. Natural mechanisms for absorption of greenhouse gases, such as 'sinks' (forests and oceans), and greenhouse gas emission reduction targets at country level through cleaner and energy-efficient technologies, and carbon trading, have been prioritised. In her assessment of the efficacy of considering gender issues in international processes aimed at developing climate change policy, Margaret Skutsch explains why these issues are not always significant at the global level. For example, apportioning blame or responsibility by sex for the cause of greenhouse gas emissions, on the basis of energy practices and consumption, is not effective. This is because most greenhouse gas emissions stem from industrial patterns of production and consumption, and require technical and scientific solutions to offset or decrease them. The role that gender issues play in global efforts is more significant in adaptation measures where

gender concerns are most stark, and where gender advocates can meaningfully contribute to developing gender-sensitive policies and initiatives.

There may be opportunities for women to acquire wider development benefits through the climate change process. Skutsch and Denton suggest that the Clean Development Mechanism (CDM) may offer new and additional opportunities for gender-redistributive initiatives. The CDM is one of the three flexible mechanisms[2] introduced in the Kyoto Protocol and the United Nations Framework Convention on Climate Change Co-operation (UNFCCC) as a co-operative instrument to promote sustainable development in developing countries, as well as cost-effective greenhouse gas mitigation. If women can access CDM funds for climate action (mitigation, adaptation, and capacity-building), then they could gain some benefits, for example by acquiring cleaner technologies for cleaner household energy usage (such as the UPESI stoves in Kenya and battery-operated lamps in Bangladesh). Such small-scale measures may not have a significant impact on greenhouse gas mitigation, but may serve to reduce respiratory illnesses, and provide an entry point for educating communities about the threats of climate change. Also, such steps could decrease women's time and labour spent in firewood collection.

At present, however, women's capacity to influence international decisions and outcomes over climate efforts remains limited. Participation in international negotiations on climate policy and interventions, such as the UNFCCC and Kyoto Protocol, is of strategic importance to women if international initiatives are to address their concerns. Capacity-building for women in this area requires equal access to education, training, and technology in developing countries, and more female professionals and male experts who have received gender training in the fields of engineering and other technical areas, who could potentially contribute to a more gender-sensitive CDM policy. It also means equipping women with the expertise to negotiate, and to conduct climate-related research relevant to their needs and interests. Strategies to overcome financial and time-investment barriers to participation in climate fora, which can deter newcomers and particularly poorer participants and women, are essential.

Including women in climate change negotiations can enhance the process, particularly where they can draw on strengths such as networking, inter-personal skills, and an ability to co-operate. Delia Villagrasa provides examples of how women were able to influence climate negotiations to some extent because of such skills. Mary Jo Larson suggests approaches that can assist disadvantaged groups in negotiating and transforming power relations. Drawing on an analysis of power relationships and climate change negotiations, she discusses how capacity-building can be a flexible and multilateral approach to sustainable development. Proactive, co-operative approaches such as those taken by the Association of Small Island States (AOSIS) for climate negotiations demonstrate how low-power groups can transform threatening systems by building alliances, developing extensive communication networks, and advocating with a united voice.

Many articles in this journal highlight how women are not 'victims' or inactive political agents. Dankelman provides examples of women organising and influencing sustainable development initiatives by lobbying for more gender-sensitive policies during the United Nations Conference on Environment and Development (UNCED) process, with some success. Women are also continuing their lobby efforts towards preparatory meetings for the World Summit on Sustainable Development (WSSD). One such organisation is the Women, Environment and Development Organisation (WEDO), which has played an important facilitating role in the WSSD consultation process and preparatory meetings, and has developed a resource book as a tool for the process.

Together, Tieho Makhabane, Irene Dankelman, and Delia Villagrasa illustrate women's positive contributions to local, national, and international processes around energy, sustainable development, and climate change issues, demonstrating how women's efforts have changed over the years. Makhabane provides an example of two energy networks, SAGEN and ENERGIA, which promote the role of women in sustainable energy development in Africa. She provides two case study examples of sustainable energy networks, and their achievement in networking around sustainable energy issues and building women's capacity.

Conclusion

This collection of articles explores some of the complex and nebulous political and socio-economic issues linked to climate change. While there is increasing consensus around the scientific hypotheses suggesting an upturn in major climatic events, policy-makers are faced with major difficulties in assessing how gender-differentiated outcomes of climatic threats may be mitigated. Firstly, the exact nature, scope, and timescale of local impacts cannot be accurately determined. Secondly, the issue is highly politicised, with major political and corporate interests at play. Thirdly, although potentially cataclysmic, the threat of climate change may not be perceived as demanding immediate attention by poor communities and countries with other, more immediate, practical concerns.

However, in many parts of the world, extreme climatic events and climatic changes are already being experienced, albeit on a smaller scale and with less frequency than can be expected in the future. Risk management is a necessary response to this reality. The articles in this journal demonstrate that gender and poverty considerations need to be included in all adaptation efforts. A better understanding of the connections between gender and poverty, the ways in which they increase vulnerability to climate hazards, and their implications for the impacts of climate change on livelihood and survival strategies, is essential. Reducing poverty requires ensuring that the poor and vulnerable have access to productive resources, land and property rights, adequate information, sound technologies, and relevant skills. All of these are crucial to natural resource management, and all represent structural constraints faced by women in many societies. These are also the kinds of areas covered in 'traditional' development agendas.

Minimising vulnerability linked to climate change impacts will require sustainable development interventions in multiple sectors (agriculture, health, employment, education, and so on) to address the incipient threats. Further research is required into the gender-differentiated impacts and vulnerabilities of climate threats, and policy discourses need to shift to accommodate the equity and sustainability implications of climate change.

Notes

1 Changes in climate occur as a result of both internal variability within the climate system and anthropogenic – or human-induced – factors. While the global climate system has always experienced natural fluctuations over time, anthropogenic factors are today creating considerable and widespread changes (IPCC 2001a).

2 The Kyoto Protocol, one the main instruments for tackling climate change, has three main mechanisms: International Emissions Trading (IET), Joint Implementation (JI), and Clean Development Mechanisms (CDM) (Humphreys 1998).

References

Humphreys, S. (1998) 'Equity in the CDM', Dakar: Enda, http//www.enda.sn/energie/cc/cdmequity.htm (last checked by author May 2002)

IPCC (2001a) *IPCC Third Assessment Report – Climate Change 2001: Summary for Policymakers*, http://www.ipcc.ch/ (last checked by author April 2002)

IPCC (2001b) *Climate Change 2001: Impacts, Adaptation and Vulnerability, Summary for Policymakers*, http://www.ipcc.ch/ (last checked by author April 2002)

Martens, P. (1998) *Health and Climate Change: Modeling the Impacts of Global Warming and Ozone Depletion*, London: Earthscan

Oxfam (2000) 'Climate Change: The Implications for Oxfam's Programme, Policies, and Advocacy', unpublished paper, Oxford: Oxfam

Climate change vulnerability, impacts, and adaptation:
why does gender matter?

Fatma Denton

Gender-related inequalities are pervasive in the developing world. Although women account for almost 80 per cent of the agricultural sector in Africa, they remain vulnerable and poor. Seventy per cent of the 1.3 billion people in the developing world living below the threshold of poverty are women. It is important that the consequences of climate change should not lead already marginalised sections of communities into further deprivation. But key development issues have been at best side-tracked, and at worst blatantly omitted, from policy debates on climate change. The threats posed by global warming have failed to impress on policy-makers the importance of placing women at the heart of their vision of sustainable development. This article argues that if climate change policy is about ensuring a sustainable future by combining development and environment issues, it must take into account the interests of all stakeholders. The Global Environment Facility and the Clean Development Mechanism of the Kyoto Protocol can play a role in ensuring sustainable development, provided they are implemented in a way that does not disadvantage women and the poor.

What have unequal power between women and men, and global inequality, got to do with an environmental crisis as monumental as the possible negative impacts of climate change – which are predicted to have far-reaching implications for women and men? The answer to this question is not immediately obvious. Hurricanes, floods, and other incidents related to climate change affect whole communities, and should presumably therefore affect the lives of women and men equally. Moreover, ecosystems and extreme climate events are oblivious to boundaries. The planet is a global concern incorporating a multitude of ecosystems, peoples, and cultures. As such, it requires collective input in its management, protection, and ultimately, its sustainability. Yet climate negotiations could be seen as a parody of an unequal world economy, in which men, and the bigger nations, get to define the basis on which they participate and contribute to the reduction of growing environmental problems, while women, and smaller and poorer countries, look in from the outside, with virtually no power to change or influence the scope of the discussions.

More than a decade since it began, the climate debate continues to be fraught with difficulties. Protagonists have gradually awakened to the fact that the underlying capitalist and market forces are too important to ignore. The debate has swayed from an initial commitment to greenhouse gas mitigation to trying to get recalcitrant countries such as the USA to toe the line and ratify the Kyoto Protocol. Climate change negotiations such as those leading to the Kyoto Protocol reflect Northern priorities and interests. Issues facing people living in poverty – such as the question of how they can adapt to climate changes – have been side-tracked or omitted.

Whilst delegates dwell on the 'shoulds' and 'woulds' of the Kyoto agreement,[1] poorer communities in Mozambique and other developing countries know that it will take more than semantics to reverse some of the most catastrophic outcomes of climate variability and environmental degradation. Most less-developed countries (LDCs) feel that their need for adaptation strategies has not been met or received sufficient attention. In the interim, whilst international agencies haggle over who is best able to oversee adaptation projects, rich industrialised countries keep a steady eye on the costs. Endless discussions ensue over what some see as a miserly adaptation fund,[2] but which others, in the North, consider to be generous. Ordinary people in rural Africa and other parts of the developing world are left to find their own ways of cultivating their land and resisting further environmental degradation, as ecosystems become more fragile and affected by climate variability.

Climate change is likely to accentuate the gaps between the world's rich and poor. It is widely accepted that women in developing countries constitute one of the poorest and most disadvantaged groups in society. A number of human practices are likely to worsen the current scenario of environmental degradation, and increase the build-up of greenhouse gas emissions in the atmosphere. Among these are energy intensity, deforestation, burning of vegetation, population growth, and, ultimately, economic growth.

Women's contribution and participation can help or hinder in all the above scenarios. It has been well documented that rural women in particular play a key role in environmental and natural resource management. Women's active involvement in agriculture, and their dependence on biomass energy, makes them key stakeholders in effective environmental management. Hence, women and their livelihoods activities are particularly vulnerable to the risks posed by environmental depletion (Denton 2001). The need to diversify energy resources and introduce alternative fuels for household use constitutes an essential part of adaptation strategies.

Taking preventive measures well in advance has more benefits than reacting to unexpected catastrophes. To plan these, it is important to consider sectors of production, such as agriculture and fisheries, in terms of the division of labour between women and men, and to identify the different degrees of vulnerability of women and men to the negative effects of climate events. Building this analysis will require more research, but this would enable policy-makers to put measures in place to combat environmental degradation, with the aim of minimising the vulnerability of the women and men affected by them. In planning such measures, much can also be learned from existing mechanisms for drought control by regional groupings such as the Permanent Inter-States Committee for Drought Control in the Sahel (CILSS). These help to build resilience, identify warning signs to give advance warning of problems, and create a sense of preparedness among women and men.

Ignoring women's contribution to environmental resource management

Women's absence from decision-making processes

Women are patently absent from the climate change decision-making process. The climate debate has not sought to address the existing marginalisation of women, nor their need to be integrated in environmental policies. Nor have the immediacy of global warming, the magnitude of such a phenomenon, or even extreme events such as the floods in Mozambique, succeeded in impressing on decision-makers the importance of placing women at the heart of sustainable development.

Increasing participation of women in UNFCCC bodies and the Kyoto Protocol is essential if policies are to promote rather than hamper gender equity. At the Seventh Conference of Parties (CoP7)[3] under the United Nations Framework Convention on Climate Change (UNFCCC) held in Marrakech, Morocco from 29th October-10th November 2001, the delegate from Samoa called for a more equitable representation of women within the organisational and decision-making structure of the UNFCCC (UNFCCC 2001). Consequently, and as a result of other dissenting voices, the CoP7 thought it necessary to improve the representation and participation of women in bodies established under the UNFCCC and the Kyoto Protocol. However, ensuring women's participation in these debates will not guarantee that the many issues faced by women in poverty will be addressed.

Poverty is linked in a complex way to exclusion and marginalisation, and this results in the absence of people living in poverty, and a lack of analysis of the issues they face, in macro-economic policy-making. Poverty leads to poor women and men being unable to make choices that might improve their socio-economic conditions, and protect natural resources. Hence, reducing poverty must be about ensuring that the poor have access to reproductive resources, control over and access to fertile land, adequate information, sound technologies, relevant skills, adequate sanitation, good irrigation strategies, and access to clean water. All of these are crucial to resource management and conservation of biodiversity.

In many cases, international debates have sidelined the priorities of the poor – particularly women – in favour of 'highbrow' discussions on technicalities including fungibility and certified emissions reductions. In addition, decision-makers, policy-makers, researchers, and development planners alike claim to represent the interests of 'the people', but use language that is seldom understood by the very people they intend to serve or represent. If addressing the negative effects of climate change is a prerequisite to sustainable development, then it is imperative that the debate is given a people's perspective.

The United Nations Conference on Population and Environment recognised the value of women in natural resource management, and their intrinsic importance was reflected in the Agenda 21 documentation. However, women are for the most part not well-represented in environmental policy formulation. The climate debate is perpetuating the under-valuation and misunderstanding of women's contribution to environmental management. While a great deal of lip-service has been paid to women's indigenous knowledge of environmental management and soil preservation, little is being done to integrate this local knowledge into mainstream policy. The African Women Leaders in Agriculture and Environment (AWLAE) came together partly to ensure that women's contributions in agriculture do not go unrecognised by policy-makers, researchers, and development planners.

Mainstreaming gender perspectives within conservation and natural resource management

As highlighted earlier, poor women are generally on the receiving end of the effects of increasing environmental degradation and depletion of natural resources, because of their involvement in, and reliance on, livelihoods activities which depend directly on the natural environment. For example, environmental degradation surrounding rural communities may increase the distances that women have to walk in search of clean water and firewood in order to perform their daily household chores.

The development sector as a whole, from energy to agriculture, seems to 'mainstream' gender issues as an afterthought. In addition, mainstreaming is done in small doses, with considerable time elapsing between times at which different development sectors adopt a gender analysis. In such an environment, policies which evolve from a gender perspective tend not only to be minimal and unenforced, but are also created in isolation from other key development sectors, and therefore offer little potential for poor rural women to optimise their skills and make significant gains.

For example, the 1980s witnessed a new form of green revolution in Africa, including within the Sahelian countries. Here, environmental management is hampered by the existence of fragile ecosystems and natural hazards such as drought and desertification. Policy-makers have particularly advocated reforestation projects. However, in addition to the wealth of services provided by forests to the rural poor, forests are known to be huge depositories of methane, nitrous oxide, and carbon dioxide.

There is little logic in involving women in environmental conservation and tree-planting schemes when only a small percentage of women have control over land. Development policy has still not addressed the structural constraints that curtail women's access to control and ownership over resources. Nor has it been able to ensure that women's expertise in land or resource management is recognised, or their effort compensated, through ownership of that land or resource.

Of course, it should be pointed out that while some analysts have emphasised the importance of women owning their own land as if land ownership in itself is a guarantee of economic prosperity, this is not the case in Africa. The majority of farmers in Africa are poor and getting poorer, mainly due to declining terms of trade, their inability to control prices on the world market, and ineffective national policies that seek to make profits from agricultural earnings. While the relationship of land ownership to productivity, and its potential for economic development, cannot be underestimated, efforts should be made to understand and ultimately address the causes of environmental degradation from a holistic standpoint, rather than conflating the single issue of lack of land ownership with women's growing poverty. To return to the example of tree-planting projects, scant attention tends to be given to other gender issues important to the success of these interventions. For instance, while women have often been used as promotional agents for tree-planting schemes, little attention has been given to the primacy of education in ensuring good resource management and environmental conservation. Minimal work, if any, is undertaken to hone women's indigenous knowledge and expertise. Through close interactions with forest and other ecosystems over many years, women have developed a wealth of indigenous knowledge of plants and their medicinal value. Sadly, this component of local knowledge has not widely been tapped into by policy-makers, and could be lost if it is not used (Agarwal 1992).

Peeling through different layers of vulnerability: potential impacts of climate change

Climate change is a threat to human security in general. A key priority in the current climate-change discussion is to ensure that decision-makers and key stakeholders alike understand the different types of vulnerability to climate change that women and men face, and their gendered implications.

According to Robert Watson, chair of the IPCC, 'vulnerability' can be defined as:

14

'...the extent to which the natural or social system is susceptible to sustaining damage from climate change and is a function of the magnitude of climate change, the sensitivity of the system to changes in climate. Hence, a highly vulnerable system is one that is highly sensitive to modest changes and one for which the ability to adapt is severely constrained.' (Olmos 2001, 3)

Vulnerability and adaptation to the adverse impacts of climate change are the most crucial environmental concerns of many developing countries, and particularly of those in the Sahel region. Different regions and countries face differing levels of vulnerability to climate changes, with commensurate differences in the vulnerability of communities and services affected by the changes. It is envisaged that climate change will affect a whole host of areas, including habitats, wildlife, terrestrial and aquatic ecosystems, and hence the production of goods and services which depend on these natural resources. Climate change will result in severe adverse changes in soils, arid-lands, coastal zones, and tropical and boreal forests (Downing *et al.* 2000). In addition, wetlands and vulnerable species would be under severe threat.

The greenhouse gas emissions of African countries are insignificant in global terms; the major sources of emissions that exist arise from land-use changes and deforestation. However, West Africa, and particularly the Sahel, is one of the most vulnerable areas to climate change, due to its propensity to drought and desertification, and its dependence on subsistence agriculture. Vulnerability to poor rainfall is the most striking feature of the Sahelian countries in West Africa (Denton *et al.* 2002). Climatic uncertainty and rapid population growth mean that the Sahel region is continually under threat of a breakdown in natural balances. This poses a threat to Sahelian peoples' access to the basic human rights of food security and

access to safe water, a sustainable livelihood, minimal exposure to health hazards, and education.

While links between human security and environmental change may not lend themselves very easily to concrete examples, land degradation is clearly connected to economic and food insecurity. Africa, as a continent, is highly vulnerable to the effects of desertification and desiccation, since 85 per cent of its water is used for agricultural purposes, and this agriculture ensures food subsistence and food security for millions of people. It is estimated that agricultural yields could reduce by up to one-tenth in some situations (UNEP/IUC 2001a). Already, many women have to take short-cuts in food preparation as a result of energy poverty, and have to resort to less nutritive meals in order to compensate for increasing fuel shortages. Consequently, they have to reduce the energy expended in their quest for fuel and fodder.

Climate change is also predicted to exacerbate existing shortfalls in water resources. The large river basins of the Niger, Senegal, and Lake Chad have experienced total water decrease of between 40-60 per cent (UNEP/IUC 2001b). More water shortages would cause further difficulties for women in health and sanitation, as rural women in Africa live mainly in water-stressed areas, and are already bearing the brunt of water shortages. Increased water contamination as a result of human activities increases vulnerability to diseases such as trachoma and scabies. Rivers and dams have become major dumping-sites for agricultural and industrial waste – yet these are also invariably the only form of water that women and children have at their disposal. In West Africa, changes in seasons and climatic conditions mean that women and children have to use water supplies from dirty ponds, as these tend to dry up during the dry season, increasing their exposure to water-borne diseases. Further, the increased time taken to fetch water may entail that young female household members are additionally required to help with household duties, increasing the likelihood of their missing school.

Health problems, such as cardio-vascular and other respiratory diseases, are likely to increase as a result of climate change, and increased temperatures could entail loss of lives. Although women in most countries have a longer average life expectancy than men, the quality of women's health is low compared to that of men in their households and communities. While health threats related to global warming linger, women are faced with more immediate health risks than men, due to their role in the gender division of labour. As fuel-collectors and cooks, they face respiratory problems caused by indoor pollution due to their direct contact with traditional fuels. As water collectors, they face high exposure to malaria, endemic in many parts of Africa. Women also share with men the risk of contracting water-borne diseases, such as dysentery, diarrhoea, and cholera. The latter are widespread in many parts of Africa, especially in extremely deprived areas, where the availability of clean drinking water is non-existent. They are likely to suffer increased, nutritional problems due to their low income status. Increased poverty and food insecurity will also cause problems relating to anaemia, common in pregnant women, and health risks accentuated by paucity of resources in both pre- and post-natal care.

Adaptive capacities and mechanisms

The Global Environment Facility (GEF) is the financial mechanism of the UNFCCC. It was created to assist non-Annex I Parties[4] in the management of the global environmental commons, by providing financial assistance to developing countries to comply with their obligations, as stipulated under the Convention. The GEF was established to forge international co-operation and to finance actions to address four threats to the global environment: biodiversity loss, climate change, degradation of international waters, and ozone depletion. Measures to halt land degradation are also eligible for GEF funding.

The Clean Development Mechanism (CDM) is a mechanism in the Kyoto Protocol conceived to allow for, and ultimately address, divergent objectives and priorities between the North and the South. It is a bilateral agreement between an industrialised country that must reduce its greenhouse gas emissions under the Convention, and a developing country. Under the CDM, industrialised countries invest in projects that increase economic productivity and may reduce local environmental problems in developing countries. CDM projects will produce commodities for a market in emissions reduction credits. Energy-efficiency projects in a non-Annex I country will be a more effective method of emissions reduction than the production of a similar system in an Annex I country. If the CDM is properly managed, it could allow non-Annex I countries to orient their goals towards development paradigms that are inseparable from sustainable development trajectories. The CDM encourages developing countries to integrate the concept of sustainability into their overall development priorities. CDM has huge potential, yet poses an undeniable challenge: how can we allow developing countries to outline and achieve their hopes of sustainability while enabling Northern countries to reach their objectives of emissions reductions?

For example, CDM projects could aim to bring about greater sustainability by promoting cleaner fuels. Projects promoting the use of improved stoves were popular in the 1970s and 1980s as a way of reducing deforestation and helping poorer households move up the energy ladder. In a

UNDP case-study on generating energy-related opportunities for women, rural women in Bangladesh who have been able to use battery-operated lamps have made significant reductions in indoor pollution, a problem which primarily affects women and children through cooking with firewood and biomass resources in confined spaces. The large-scale use of batteries instead of polluting energy resources such as kerosene have the potential to reduce greenhouse gas emissions (Khan 2001). Likewise, Upesi stoves (cleaner, more fuel-efficient, faster cooking stoves, designed and disseminated in Kenya) have not only created savings in fuel and time for women, but have also created a greater awareness that energy and soil conservation are central to environmental sustainability (Njenga 2001).

Some authors believe that the CDM might constitute a double-edged sword, in that it will benefit those countries and individuals best able to take advantage of it. CDM projects would best serve countries that have the necessary capacity and institutional arrangements to take advantage of energy projects. Least-developed countries are least likely to gain from these initiatives, because their economies are weak, their institutions feeble, and their human resources minimal. By the same token, although increased prominence in biomass energy could open up opportunities for entrepreneurs and attract businesses, the opportunities would go to people who are most able to take advantage of these (Kammen 1995). Structural constraints such as lack of education and entrepreneurial skills, and cultural restrictions, tend to inhibit women's efforts in entrepreneurial activities (Colletah 2000). CDM projects should support existing initiatives like the two outlined earlier, and strengthen the capacity of those involved in them. Too often, projects are designed without prior consultation with women, and environmental and other benefits end when the

project finishes. Governments and the private sector need to find ways of identifying potential project portfolios and of making sure that the main stakeholders are included at all stages of project development.

CDM has great potential to create and support energy infrastructures that could benefit large populations within the African continent. It could also enhance regional and sub-regional co-operation. Such an opportunity will be immensely beneficial in the areas of transport, housing, and electricity provision. Projects could be transposed and emulated from one country to another, in a North-South-South triangle.

Socio-economic sustainability and the equity dimension

It is quite clear that building the necessary capacity to enable the least-developed countries to cope with extreme events, such as major flooding, is of absolute importance. It is important to note that there is no scientific evidence of any direct causal linkage between specific extreme climate events and climate change. However, because extreme events are liable to have considerable impact on human beings in future, because there is a high probability of climate change being linked with an increase in extreme climate events, and because the effects of day-to-day climate change will be similar to these very extreme events, it is judicious to monitor extreme events and learn to cope with them.

Promoting resource management in a sustainable manner, against a background of globalised economic change and economic austerity brought about by structural adjustment policies, poses many different challenges. Ineffective, gender-blind policies, and entrenched patriarchal traditions, are to a large extent responsible for the numerous constraints women face in transport, access to land, income, and other resources, agricultural practices, education, health services, credit facilities, and a litany of structural, technological, and cultural barriers.

Gender inequalities continue to exist in terms of access to land, control over resources, ability to command and access paid labour, capacity, and strategies for income diversification, as well as time spent on agricultural or forestry-based activities. States must promote gender-sensitive policies with regard to land tenure. Legal and structural barriers need to be overturned in order to encourage and promote equitable access to land and resources, boost productivity, and manage environmental and soil conservation. Landlessness is forcing women to adopt other income-generation activities to complement their earnings. Commercialisation of agriculture also means that women need to find alternative sources of income. For women who lack the education to filter through the different channels of bureaucracy to take advantage of credit facilities and employment prospects, communal resources may be all they have at their disposal.

Unequal power relations between women and men lead to their differential access to environmental resources and opportunities for income diversification, entailing that environmental vulnerability, and indeed security, affect women and men differently. The increasing movement of male migrants to 'greener pastures' has tended to compound the poverty that many rural women have to contend with. Policies to curb migratory flows through rural development, provide markets for local products, diversify livelihoods, and promote good business initiatives, are critical for gender equality and resource management.

Conclusion

There is enough evidence to show that women are at the centre of sustainable development, and that ensuring greater gender equalities in all sectors would mean that society as a whole will benefit. Yet mainstreaming gender issues into debates on climate change and sustainable development is happening piecemeal, extremely slowly, with varying degrees of success, and often as an afterthought. This is made more complex by women's lack of participation in decision-making at all levels, and the fact that the climate debate so far has made little effort to package the issues in a way that ordinary people can even understand, let alone participate in.

Poor economies find it difficult to meet sustainability criteria, whether they be social, economic, or technological. Power dynamics characterise the relationships between richer and poorer nations, and these have gendered implications. If poorer nations are finding it difficult to get richer nations to meet their obligations and work towards climate mitigation, poor women have an even bigger problem in promoting their agenda. If smaller and poorer nations have difficulties in mounting the necessary infrastructure to take advantage of CDM projects, poorer women have even fewer means and less scope to diversify their livelihood and look after their families.

Women are usually left subsidising economies through the energy that is expended in the agricultural, forestry, and industrial sectors. Environmental management is highly gendered, therefore conservation should also take into account gendered divisions of labour in order to achieve greater equity. Climate change is treated as a scientific event, yet its implications will have far-reaching human dimensions. So far, it has been given an economic emphasis, resulting in polemical debates and power dynamics. Thus, rich nations continue to dodge the real issues, while

18

smaller or less powerful states ensure that they stay in line and accept whatever concession they deem necessary to keep the Kyoto agreement together. Within this debate of pure economics, technicalities, and muscle flexing, the gendered implications of global warming are totally ignored.

Adaptation funds should be more than merely a 'paper victory' (Najam 2001), but should be put into operation, to allow marginalised sections of communities to move out of poverty. CDM projects should find a way of building the capacity of poor men and women, to ensure that self-reliance is attained and that future generations are spared the task of cleaning up an environmental mess they did not generate. It is certainly not enough to look at adaptation from a merely welfaristic perspective, as this would ignore the roles played by different stakeholders, and would consequently entrench and accentuate the very inequalities it is seeking to blur. Climate equity is about ensuring that some voices are not muffled at the expense of the more vociferous and powerful ones. Human adaptation to climate change is a very practical concern. Discussions must not only reflect a cross-section of society, but should also aim to ensure that people with low resilience are given the necessary tools to adapt and ultimately sustain themselves.

Climate change policy is based on a 'survival of the fittest' philosophy. There is very little either in the Kyoto Protocol or the Convention to protect the migrant farmer faced with drier soils, heat stress, and lower productivity. Equity matters, in that climate change, through its policies, should not open the floodgates of consumerism in the North if this is going to be at the expense of Southern fishermen or agriculturalists faced with problems relating to sea-level rise and loss of livelihoods.

Women are already paying huge prices for globalisation, economic depression, and environmental degradation. Climate change is likely to worsen their already precarious situation, and leave them even more vulnerable. More efforts should be made to give climate negotiations a people-centred approach, and to give women their rightful place within the sustainable development circle. Women have taken a keen interest in environmental resource management, and have generated a great deal of wealth in terms of indigenous environmental knowledge, which needs to have its value to medicine and other fields acknowledged. Policy-makers have continued to make specific reference to this knowledge, but very little effort is made to utilise it, or to make using this knowledge a fundamental part of main-stream policy. This would build women's capacity and give them greater scope to utilise their potential. Unless efforts are taken to incorporate indigenous knowledge into mainstream policy, it will continue to be a case of standing knee-deep in the river and dying of thirst.

Adaptation projects should not only serve as an inventory to 'peel off' the different layers of vulnerability of different areas and people, but also to ensure that the relevant incentives are provided so that women and men can complement their efforts, and build a sustainable future through using their comparative advantages. Women's invisibility, and the diminution of their roles, has been cited as one of the reasons why gender asymmetries have been so stark, particularly in the developing world. Yet changes at micro-level, however welcome, do not even begin to scratch the surface of much-needed policies at macro-level. Climate change policies should enable developing countries to get it right the first time, through the use of smart and clean technologies whilst avoiding greenhouse gas emissions. The 'no regret' principle[5] is today a fundamental tenet of the climate negotiating policy. Policy-makers within the climate debate must ensure that this tenet is extended to marginalised groups by giving women and men an opportunity to build their capacity, lower their vulnerability, and diversify their sources of income.

Fatma Denton is Researcher and Project Co-ordinator in the Energy Programme at Enda Tiers Monde. Contact: Energie Programme, 54 Rue Carnot, Dakar, Senegal. E-mail: energy2@enda.sn or fatma@africainformation.net

Notes

1. A lot of time and energy was spent in analysing the document on compliance and debating whether the text should be left as it is or changed with new emphasis on words such as 'should' and 'would' that other parties felt uncomfortable with.
2. At CoP7 (the Seventh Conference of Parties) , it was decided that in order to strengthen existing national climate change secretariats, build capacity in language and negotiating skills among Least Developed Countries (LDCs), and facilitate the process of National Adaptation Programmes of Action (NAPAs), some funds needed to be set aside.
3. The Conference of Parties comprises 185 members and is made up of the majority of world states who act as members. It is the supreme body of the Climate Change Convention and it meets once a year.
4. Non-Annex I countries are developing nations that do not have to agree to the Kyoto Protocol emissions caps.
5. The 'no regret principle' means greenhouse gas emission reduction options that constitute negative net costs. This is because they generate direct or indirect benefits that are considerable enough to outweigh and offset the costs of implementing the options.

References

Agarwal, B. (1992) 'The Gender and Environment Debate: Lessons from India', *Feminist Studies* 1 (Spring)

Colletah, C. (2000) 'Culture as a barrier to rural women's entrepreneurship: experience from Zimbabwe', *Gender and Development* 8(1): 71-7

Denton, F. (2000) 'Gendered impacts of climate change. A human security dimension', *Energia News* 3(3):13-14, http://www.villagepower2000.org/africa/aem2000/presentations/ondiaye abstractenglish.doc (last checked by author May 2002)

Denton, F. (2001) 'Climate change, gender and poverty – academic babble or realpolitik?', *Bulletin Africain: Point de Vue* 14, Dakar: RABEDE

Denton, F., J.P. Thomas, and Y. Sokona (2002) *Climate Change and Sustainable Development Strategies: An Agenda for Long Term Action*, OECD Environment and Development Co-operation, Paris: OECD

Downing T.E., Y. Sokona, and J.B. Smith (2000) 'Action on Adaptation to Climate Change', presentation to the UNFCCC workshop 'Article 4.8 and 4.9 of the convention: adverse effects of climate change', Bonn, Germany, 9-11 March 2000, Oxford: Oxford Environmental Change Institute

Kammen, D. (1995) 'From energy efficiency to social utility: improved cookstoves and the small is beautiful model of development', in J. Goldemberg and T.B. Johansson (eds.), *Energy as an Instrument for Socio-Economic Development*, New York: UNDP

Khan, H.J. (2001) 'Battery-operated lamps produced by rural women', in G.V. Karlsson (ed.), *Generating Opportunities: Case Studies on Energy and Women*, New York: UNDP

Najam, A. (2001) 'Good and Bad News', in *Equity Watch – Green Future*, http://www.cseindia.org/html/cmp/climate/ew/art20011130 1/htm

Njenga, B.K. (2001) 'Upesi rural stoves project', in G.V. Karlsson (ed.), *Generating Opportunities: Case Studies on Energy and Women*, New York: UNDP

Olmos, S. (2001) 'Vulnerability and Adaptation to Climate Change: Concepts, Issues, Assessment Method Paper', *Climate Change Knowledge Network Foundation Paper*, http://www.cckn.net (last checked by author May 2002)

UN Environment Programme Information Unit for Conventions (UNEP/IUC) (2001a) 'Human settlements, energy and industry', *Climate Change Information Kit*, Sheet 15, Geneva: UNEP Information Unit for Conventions, http://www.unep.ch/conventions/

UN Environment Programme Information Unit for Conventions (UNEP/UC) (2001b) 'Water Resources', *Climate Change Information Kit*, Sheet 13, Geneva: UNEP Information Unit for Conventions, http://www.unep.ch/conventions/

UNFCCC (2001) 'Improving the participation of women in the representation of Parties in bodies established under the United Nations Framework Convention on Climate change', draft decision proposed by the President, Conference of the Parties, Other Matters, Seventh session, Marrakech, 29 October–9 November 2001, Agenda item 13, France, UNFCCC

Climate change:
learning from gender analysis and women's experiences of organising for sustainable development

Irene Dankelman

This article argues that climate change not only requires major technological solutions, but also has political and socio-economic aspects with implications for development policy and practice. Questions of globalisation, equity, and the distribution of welfare and power underlie many of its manifestations, and its impacts are not only severe, but also unevenly distributed. There are some clear connections, both positive and negative, between gender and the environment. This paper explores these linkages, which help to illustrate the actual and potential relationships between gender and climate change, and the gender-specific implications of climate change. It also provides examples of women organising for change around sustainable development issues in the build-up to the World Summit on Sustainable Development (WSSD), and demonstrates how women's participation can translate into more gender-sensitive outcomes.

Climate change[1] is not an occurrence in the distant future, but a phenomenon that is taking place now. The Intergovernmental Panel on Climate Change (IPCC), established by the United Nations in 1988 to gather data and develop knowledge about climate change, presented its Third Assessment Report (Working Group I) in 2001. It concluded that global warming was a reality, and that there was new and stronger evidence that most of the warming observed over the last 50 years was attributable to human activities. An increasing body of observations gives a picture of a warming world and other changes in the climate system. These changes are presented in Box 1.

The human dimensions of climate change

It is clear that human interventions are largely causing these changes in the global and local climate systems (IPCC 2001b). Emissions of greenhouse gases due to human activities continue to alter the atmosphere. Higher concentrations of greenhouse gases, such as carbon dioxide and methane, warm the earth's surface. About three-quarters of the human emissions of carbon dioxide are due to fossil fuel burning; the rest is due to land-use change, especially deforestation. More than half of the increase in methane emissions is caused by human activities, such as use of fossil fuels, cattle, rice agriculture, and landfills.

This raises the question of who is responsible for causing climate changes, and what are the main effects of these changes on different sectors of the population and environment. A response to this demonstrates why climate change is a development issue, and proves that it has a political character. Historically, industrialised nations have emitted 80 per cent of their greenhouse gases due to their rapid industrialisation. Social and economic developments in the industrialised countries took place at the expense of the

---

Box 1: *Observed and projected climate-related changes*

- The earth's average surface temperature has increased by about 0.6°C over the 20th century. This results in higher maximum and minimum temperatures, and more hot days.
- Snow cover and ice extent have decreased.
- Global average sea level has risen, and ocean heat content has increased.
- More frequent precipitation, linked to warming, will cause increased flooding in some parts of Asia and Africa.
- The frequency and intensity of droughts is projected to increase.

Source: IPPC 2001a

---

colonised world. Most of the greenhouse gases today are emitted into the atmosphere by these same industrialised countries, which consume large quantities of fossil fuels. At the same time, industrialised countries including the USA, Canada, Japan, and Australia are the main blocks to progress in the UN climate change negotiations (CJN 2001a).

Who is most affected?

Ironically, climate change effects and related disasters have occurred predominantly in the developing world: in 1998 the melting of snow in China and India caused 5550 deaths. Typhoons, cyclones, and hurricanes in the Philippines, Bangladesh, and the Caribbean and Central America caused 15,800 deaths. In 1999, 50,000 people died in Venezuela due to heavy rains and mud floods. In 2000, a major disaster occurred in Mozambique when rains, floods, and cyclones affected 800,000 people, caused 700 deaths, made 250,000 homeless, increased the incidence of malaria and other tropical diseases, and impacted negatively on food production. People with low incomes affected by such disasters have very little to fall back on: they lack insurance, savings, or adequate social welfare structures to cope with such dramatic events. As a result, they suffer death, injury, illness, become homeless, and are forced to seek refuge in other areas or countries. This contributes to a rising

number of environmental refugees, who are still not recognised by the UN (CJN 2001).

Projected outcomes for livelihoods and human security

The IPPC Second Working Group Report (IPPC 2001b) concludes that recent regional climate changes, particularly temperature increases, have already affected many physical and biological systems. There has been increasing frequency of floods and droughts in some areas. Projected adverse impacts on livelihoods and human security include:

- a general reduction in potential crop yields in most tropical and sub-tropical regions, posing a major risk to food security;
- decreased water availability for populations in many water-scarce regions, particularly in the sub-tropics;
- an increase in the number of people exposed to vector-borne diseases such as malaria, and water-borne diseases like cholera, and an increase in heat-stress mortality;
- a widespread increase in the risk of flooding for tens of millions of inhabitants of human settlements as a result of increased heavy rainfall and sea-level rises.

The IPPC concludes that those with the least resources are the most vulnerable to the negative effects of climate change, and have the least capacity to adapt to these effects. Vulnerability to the negative effects of climate change of human populations and natural systems differs substantially across regions and populations within regions. Populations inhabiting small islands and low-lying coastal areas are at particular risk of severe social and economic problems arising from sea-level rise and storm surges (IPPC 2001b).

This indicates that there are major disparities between those who cause climate change, and those who are affected by it. The impacts of climate change will fall disproportionately upon developing countries and poor people within all countries, and will exacerbate inequalities in health and access to adequate food, clean water, and other resources. As the Climate Justice Network concludes, 'There is certainly an environmental justice aspect to climate change; and it is necessary to see the links between the environmental issue of climate change and social injustices, like racism and economic inequity.' (CJN 2001a, 1) Issues of unequal distribution of welfare and power are behind the main causes, manifestations, and effects of climate change.

Gender analysis and the physical environment

Is there a gender dimension to climate change? IPPC concludes that, 'Climate change impacts will be differently distributed among different regions, generations, age classes, income groups, occupations, and genders.'(IPPC 2001b) In order to determine the gender dimensions of climate change, it is helpful to consider studies of the relationship between gender and the environment.

Since the mid 1980s, several studies have indicated that the relationship between communities and their physical environment is not gender-neutral. Much emphasis has been placed on the fact that rural women, in developing countries in particular, interact more directly with their environment, and are disproportionately adversely affected by environmental degradation. The Centre for Science and Environment (CSE), based in New Delhi (India), argued in their *The State of India's Environment Report* in 1985, 'Probably no other group is more affected by environmental destruction than poor village women. Every dawn brings with it a long march in search of fuel, fodder and water... As ecological conditions worsen, the long march becomes even longer and more tiresome. Caught between poverty and environmental destruction, poor rural women in India could well be reaching the limits of physical endurance.' (CSE 1985, 172)

Several publications, illustrated with many case studies, have since described the different roles that women have in the management and use of land, water, energy, and biodiversity. According to feminist analyses such as that of Esther Boserup (1989), it is actually 'woman-the-gatherer' and not 'man-the-hunter' who was traditionally a source of sustainable food supply. It has also been stressed that women play a major role in actions to safeguard the environment, and therewith their communities' livelihoods and survival. This is not a new phenomenon. In the 18th century, women under the leadership of Amrita Devi were actively involved in an environmental struggle for survival in Rajasthan, India (Shiva 1998). When, in the 1970s, Cape Verde was struck with severe droughts, a women's organisation, Açao Democrática Feminina Gaúcha – which had originally focused on social and educational issues – put environmental issues high on its agenda. It has now become Friends of the Earth Brazil (Dankelman and Davidson 1988).

In her article on ecological transitions and the changing context of women's work, Geeta Menon (1991) describes work as the active, labour-based interaction of human beings with the material world. Historically, this interaction has been intricately based on the natural environment in which human populations survived. Many traditional economies were founded on a gender division of labour in which women typically had primary responsibility for certain areas of resource management. This has meant that women's connection to the environment has largely been rooted in their work.

Writers like Dankelman and Davidson (1988), and Kelkar and Nathan (1991), argued that it is wrong to talk about women as one homogeneous group, because of the vast economic, cultural, and social differences between women. Differentiating factors such as class and caste, kinship, age, nationality, and socio-cultural group are important variables. Analysing these differences is as crucial as looking into the differences between women and men (Kelkar and Nathan 1991).

Greater insight into the problem was gained with the perception that it is not enough to look at the position of women and the environment in isolation. Power relations between both sexes are determining factors.

Access to, and effective control over, natural resources such as land, water, and forests, are important indicators of gender positions. The use and management of these resources, as well as decision-making at micro-, meso-, and macro-levels, are gender-differentiated. It might be clear that if the quality or quantity of the resources upon which managers depend are affected, this also affects their work, effort, and the energy which is needed for that management, and thus limits their other development options. Not only is control over resources, re-distribution of roles and tasks, and a shift in stereotypes needed to improve women's situations (compared to those of men), but these changes are also required to sustain a healthier and more productive environment.

The relationships between women and their environments are not always positive. In the policy documents 'Gender and Environment: A Delicate Balance Between Profit and Loss', and 'Rights of Women to Natural Resources, Land and Water' (NEDA 1997a and b), it is argued that differentiation on the basis of gender is of crucial importance to an analysis of how environmental measures and changes affect gender relationships, and how changes in the relative position and status of women and men have an impact on their livelihoods. Measures that appear to have a positive short-term effect from a gender perspective might also be positive in the long run for sustainable development. But this is not always the case, for example, where income-generating activities for women demand a high input of local energy resources.

Is climate change gender neutral?

Climate change is often seen as a technical problem, requiring technical solutions. But in fact there are many social and political aspects to this complex issue. Similarly, it is often argued that climate change is gender-neutral – that is, that it affects women and men in the same ways. Yet, in many cases, communities interact with their physical environment in a gender-differentiated way.

We need to look specifically at the gendered aspects of climate change. In this, we can draw on studies conducted on gender-specific aspects of disaster prevention and mitigation, such as those presented during the Expert Group Meeting on 'Environmental Management and the Mitigation of Natural Disasters: a Gender Perspective', organised by the United Nations Division for the Advancement of Women (DAW) and the Inter-Agency Secretariat of the International Strategy for Disaster Reduction (UN/ISDR), held in November 2001 in Ankara. Case studies from different regions in the world presented at that meeting showed how women have different positions and assume different roles to men in the prevention and mitigation of natural disasters (UN DAW 2001).

Below, I set out several areas where gender roles and relations interact with climate change causes and impacts to varying extents (adapted from Wamukonya and Skutsch 2001).

1. Gender-specific resource-use patterns that can degrade the environment

Although women are the main energy managers in many households in the South, it is often male-dominated organisations that make decisions affecting energy policies and programmes. The power and petroleum sectors in North and South are male-dominated, and the majority of the funding organisations are also male-dominated. With regard to usage of energy resources, Wamukonya and Skutsch (2001) argue that the gender distribution of usage of the services of these industries is changing, with women using cars and making consumer decisions more than has previously been the case.

2. Gender-specific effects of climate change

Women's status and activities make them experience poverty differently to men, and they are often more vulnerable than men to climate change and its effects. Therefore, there is certainly a gender dimension to climate change (Denton 2000). It is not only large-scale disasters that affect women's lives and livelihood. Other less dramatic problems might occur as a result of a heavy rainy season or a drought that would have a negative impact on women's daily roles and tasks, and thereby increase their burdens.

3. Gendered aspects of mitigation and adaptation

Several disaster-related studies have argued that there is a gender dimension to disaster mitigation and environmental management (see Kumar-Range 2001 for an example). Women often cope with disasters in different ways to men. For example, a case study in the *charlands* (pieces of land resulting from the accretion of silt in river channels, which are very flood-prone) of Bangladesh showed that women's indigenous knowledge and practice of environmental management play a crucial role in the management of these lands, but that their contribution often goes unnoticed. Their technological ability to cope with the changing circumstances is demonstrated as they carry out a number of innovations and adaptations, which are generally embedded in their daily lives (Chowdhury 2001).

4. Gender and decision-making on climate change

As we have seen, women play only a limited role as producers in the energy sector or in energy policies. During the climate change negotiations, almost no attention has been paid to the need to involve women, or gender aspects, fully in the deliberations. In Marrakech, during CoP7 (Conference of Parties) 2001, the delegation from Samoa presented a resolution that called for more equal participation of women in the negotiations. Minister Jan Pronk, chairman of the The Hague CoP6 meeting (November 2000), stressed, '...the widest participation in the process of promoting and co-operating in education, training and public awareness related to climate change is crucial. In developing country households women are often the primary providers and users of energy. Therefore, the participation of women and women's organisations is crucial.' (Wamukonya and Skutsch 2001, 1) Although the percentage of women ministers of the environment has certainly increased, in CoP6 only 20 per cent of the delegates were women, and in the economic sector – where many of the relevant climate change decisions are made – this number might even be less. Good practice is easily followed. Since the Netherlands delegation was successfully led by a woman in CoP6, other countries followed in CoP7.

5. Human capacity

Wamukonya and Skutsch (2001) argue that, while capacity-building is a major area of focus to enable implementation of the UNFCCC and the Kyoto Protocol, there are still major gender-based inequities in access to education, training, and technology. They conclude that it is important to design gender-sensitive capacity-building programmes for mitigating climate change.

Women organising for sustainable development

There is a tendency to talk about gender aspects of climate change as if women are only victims. Many studies show, however, that women have been instrumental in organising themselves around environmental issues and sustainable development. For example, in the preparatory process for the United Nations Conference on Environment and Development (UNCED), held in Rio de Janeiro in 1992, women organised themselves in an unprecedented way. In 1991, 1500 women from 83 countries assembled in Miami in the first World Women's Congress for a Healthy Planet.

At the congress, women from every region in the world presented dramatic testimonies of their battles against ecological and economic devastation, before a tribunal of five eminent women judges. From this evidence and their own experiences, the participants in the congress developed the *Women's Action Agenda 21* (WAA21). The action agenda, which contained recommendations and actions for a healthy planet, was intended to form a blueprint for incorporating a gender dimension into local, national, and international decision-making into the next

century. It was specifically designed to promote women's active and equal participation in preparation for the UNCED, and in implementing its expected plan of action, Agenda 21 (WEDO 1992).

WAA21 proved to be an effective lobbying document for a more gender-sensitive UNCED process, and an important common point of reference and source of inspiration for women's groups worldwide. Its strength lay in the worldwide process through which it was developed, and the facts that it was based on women's own experiences, views, and visions, and that it was linked to the UNCED process. Also, the document's broad scope, and its analysis of different thematic areas, enabled its wide use and application. Another important factor was that the Director General of UNCED, Maurice Strong, was a strong supporter of women's involvement in the UNCED process, and was present at the congress. Women leaders, for example Bella Abzug, the founder of the Women's Environment and Development Organisation (WEDO),took the document to the most relevant fora.

The document's weakness lies in the fact that it was developed at one meeting of 'only' 1500 women, and not through a worldwide process of consultation, with the result that local groups and regional networks lacked a strong sense of ownership over it. Although attempts were made to include regional perspectives in region-specific supplements, the document has mainly been perceived as having a global focus. The use of the WAA21 after UNCED was hampered because there was no strong follow-up plan connected to its further implementation. However, the WAA21 has served as an important lobbying document.

At UNCED many women's organisations gathered and shared experiences and views on environment and development in the 'Women's Tent' or *Planeta Femea*. The Women's Tent was a co-production of Brazilian women's groups and WEDO.

Its main aim was to offer a physical space for women's groups and networks to host daily meetings at the NGO Forum on the themes presented in the WAA21. The tent, the largest in the Flamingo Park, was very successful in giving visibility to the energy and commitment of many female (and some male) participants towards just sustainable development. To the participants, it offered an important opportunity to network and strengthen their global efforts. It also served as a valuable stimulus for Brazilian women's organisations after UNCED.

A major problem for all NGO activities in the NGO Forum was the considerable physical distance between the forum and the official UN meetings, which made the official conference almost unreachable for most of the NGO activists. The Women's Tent was only in place for two weeks, and the Women's Action Agenda only the product of a limited representation of global women, but both formed an inspiring source for further actions in specific areas, such as on women and biodiversity (and later the establishment of 'Women in Diversity'), and in some regions (the establishment in Europe of 'Women in Europe for a Common Future').

The success of these efforts is shown by UNCED's outcome, Agenda 21. In this official document, 'women' are distinguished as one of the nine 'major groups' for implementation of Agenda 21. This recognition of the role of women in sustainable development, and the identification of specific actions to improve women's position and enhance their role, has been very helpful for the women's movement. Since 1992, many women's organisations have appealed to this specific chapter in Agenda 21 to underline their own concerns and activities. Although clear commitments were made in Agenda 21, a review of the progress on women's position in sustainable development during the decade since 1992, presented in the

'Women's Dialogue Paper for the WSSD' (see below), showed that implementation at international and national levels is still limited (UN ECOSOC/CSD 2001).

In the present process towards the World Summit on Sustainable Development (WSSD), to be held in Johannesburg in 2002, women's organisations are once again giving voice to their main concerns and visions, and sharing their experiences in almost all areas of sustainable development over the past ten years. Not only is a revised *Women's Action Agenda for a Peaceful and Healthy Planet 2015* being developed through a worldwide consultation process, but women's groups are actively participating in the preparatory meetings at the United Nations – in which WEDO plays a facilitating role. Nationally and regionally, many women's organisations have organised meetings to prepare for the WSSD meeting. A *Resource Book on Gender and Environment* has been developed by the Stakeholder Forum for a Common Future (Hemmati and Seliger 2001) as a tool for the process.

Apart from these international processes, there has been a great deal of regional and thematic organising among women for sustainable development. Since the beginning of the 1990s, networks such as the Women and Water Alliance, Diverse Women for Diversity, and ENERGIA (see the article by Tieho Makhabane in this collection), have been established. In Central America, efforts by the IUCN (International Union for Conservation of Nature) led to the strengthening of gender analysis in environmental ministries. In these areas, the participation of women is not only becoming more visible, but is also contributing to analyses of gender aspects of specific areas. Women living and working at the 'grassroots', such as those organised in GROOTS (see Resources in this volume), bring the concept of sustainable livelihoods and habitat aspects to the local, national, and international agendas. In this context, the 'engendering' of Local Agenda 21, to which many women's groups as well as local authorities have committed themselves, is an important process.

Conclusions

Although women's organisations were already active in the field of environment and development fifteen years ago, many of these efforts have now matured, and new ones have developed over the past decade. Lessons can be learned from these. Experience over the past decade has showed that it is essential to work in well-established organisations, with committed and experienced women from many different backgrounds.

WEDO runs programmes in the areas of governance and gender, economic and social justice, and sustainable development. Co-operation and networking in specific areas between local, national, regional, and international groups is very valuable. Communication between all levels, such as has been facilitated since 1992 by widespread use of information and communications technologies, is an important tool.

It is necessary to be able to connect macro-issues with the reality of people's lives, and the lives of poor rural women in particular. It is essential for women to be involved in official processes such as the multi-stakeholder dialogues in the WSSD process. Similarly, co-operation between women in official delegations can be very helpful. In this context the establishment of the network, Women Leaders on the Environment, in March 2002 in Finland, is a helpful development. One of the most inspiring benefits of the network is that women activists of different backgrounds and regions, and different ages, can work together and offer each other support.

These are just some experiences of women and their organisations active in the area of sustainable development. In the context of this article it is impossible to refer to all the hundreds of other examples of women who are active at local, national, and international levels. For many of them it is a new challenge to become active in the field of climate change and global justice. The above analysis shows that there is potential for women to work together on issues of gender and climate change.

Irene Dankelman is Co-ordinator in Sustainable Development at the University of Nijmegen, and Senior Adviser in Sustainable Development to the Women's Environment and Development Organisation (WEDO), Hatertseweg 41, 6581 KD Malden, Netherlands. Tel: 00 31 24 3564834; E-mail: irened@sci.kun.nl or irene.dankelman@antenna.nl

Notes

1 Climate change in IPCC usage refers to any change in climate over time, whether due to natural variability or as a result of human activity. This usage differs from that in the Framework Convention on Climate Change where climate change refers to a change of climate which is attributed directly or indirectly to human activity that alters the composition of the global atmosphere, and which is in addition to natural variability observed over comparable time periods.

References

Boserup, E. (1989) *Women's Role in Economic Development*, London: Earthscan

CSE (Centre for Science and Environment) (1985) *The State of India's Environment 1984-1985: The Second Citizens' Report*, New Delhi: CSE

CJN (Climate Justice Network) (2001a) *Climate Change and Colonialism*, http://www.risingtide.nl/issues/colonialism.html (last checked by author April 2002)

CJN (Climate Justice Network) (2001b) *Hidden Statistics: Environmental Refugees*, http://www.risingtide.nl (last checked by author April 2002)

Chowdhury, M. (2001) 'Women's Technological Innovations and Adaptations for Disaster Mitigation: A Case Study of Charlands in Bangladesh', paper prepared for the UNDAW/ISDR Expert Meeting on 'Environmental Management and the Mitigation of Natural Disasters: A Gender Perspective', Ankara, 6-9 November 2001

Dankelman, I. and J. Davidson (1988) *Women and Environment in the Third World: Alliance for the Future*, London: Earthscan

Denton, F. (2000) 'Gender impact of climate change: a human security dimension', *Energia News*, 3(3):13-14

Hemmati, M. and K. Seliger (eds.) (2001) 'The Stakeholder Toolkit: A Resource for Women and NGOs', UNED Forum, http://www.earthsummit2002.org/toolkits/women/index.htm (last checked by author April 2002)

IPPC (International Panel on Climate Change) (2001a) 'Summary for Policymakers', report of Working Group I of the IPCC, http://www.ipcc.ch (last checked by author April 2002)

IPPC (International Panel on Climate Change) (2001b) 'Summary for Policymakers. Climate Change 2001: Impacts, Adaptation, and Vulnerability', report of Working Group II of the IPCC, http://www.ipcc.ch (last checked by author April 2002)

Kelkar, K. and D. Nathan (1991) *Gender and Tribe: Women, Land and Forests in Jharkhand*, New Delhi: Kali for Women

Kumar-Range, S. (2001) 'Environmental Management and Disaster Risk Reduction: A Gender Perspective', paper prepared for UN DAW Expert Meeting on 'Environmental Management and the Mitigation of Natural Disasters: A Gender Perspective', Ankara, 6-9 November 2001

Menon, G. (1991) 'Ecological Transitions and the Changing Context of Women's Work in Tribal India', *Purusartha* 14: 291-314

NEDA (Netherlands Development Assistance) (1997a) 'Gender and Environment: A Delicate Balance between Profit and Loss', Working Paper on Women and Development no.1, The Hague: Ministry of Foreign Affairs

NEDA (Netherlands Development Assistance) (1997b) 'Rights of Women to Natural Resources, Land and Water', Working Paper on Women and Development no.2, The Hague: Ministry of Foreign Affairs

Shiva, V. (1998) *Staying Alive: Women, Ecology and Development,* London: Zed Books

UN DAW (United Nations Division for the Advancement of Women) (2001) *Environmental Management and the Mitigation of Natural Disasters: A Gender Perspective*, Report of the Expert Group Meeting, Ankara, Turkey, 6-9 November 2001, New York: UN DAW

UN ECOSOC/CSD (2001) 'Dialogue Paper by Women', multi-stakeholder dialogue segment of the second session of the Preparatory Commission for WSSD, New York, 28 January–8 February 2002

Wamukonya, N. and M. Skutsch (2001) 'Is there a Gender Angle to the Climate Change Negotiations?', paper prepared for ENERGIA for the CSD9 (Commission on Sustainable Develop-ment, Session 9), New York, 16-27 April 2001

WEDO (1992) 'World Women's Congress for a Healthy Planet', report of congress, 8-12 November 1991, Miami, including 'Women's Action Agenda 21' and 'Findings of the Tribunal', New York: WEDO

Protocols, treaties, and action:
the 'climate change process' viewed through gender spectacles[1]

Margaret M. Skutsch

This paper starts by assessing the extent to which gender considerations have been taken into account in the international processes concerning the development of climate change policy. Finding that there has been very little attention to gender issues, neither in the protocols and treaties nor in the debates around them, the paper goes on to consider whether there are in fact any meaningful gender considerations as regards (a) emissions of greenhouse gases, (b) vulnerability to climate change, and (c) participation in projects under climate funding. It concludes by suggesting some areas where attention to gender could improve the effectiveness of climate interventions and also benefit women.

It takes no more than a simple word-search of the UN Framework Convention for Climate Change (UNFCCC) and the Kyoto Protocol, the two most important treaties which relate to global efforts to combat climate change, to discover that the words 'gender' and 'women' are not mentioned in either. One might ask oneself whether the absence of reference to gender considerations in such documents matters at all; they are legalistic tracts designed to provide a general framework under which much more detailed plans have to be worked out. They do not mention 'poverty' or 'deprivation' either, and refer only in very general terms to social and economic development.

More alarming, perhaps, is the fact that there has been almost no attention to gender issues in the discourse around climate change, and particularly in areas where a gender factor could be anticipated to be important, for example where the effects of climate change are linked to poverty. Very little appears to have been written on the subject. A scan of a number of prominent journals dedicated to the climate issue reveals not a single article on the gender-differentiated implications of climate change in recent years.[2] An exception is the article by Denton (2000), in which, among other things, the author points out that owing to the feminisation of poverty, women in developing countries are more vulnerable to the effects of climate change than men.

Similarly, gender issues have not been widely discussed in the so called 'climate change process', that is, the debates that surround the formulation of climate change policy. At the Sixth Conference of Parties to the UNFCCC meeting (CoP6) in The Hague in November 2000, the topic was hardly mentioned, although the Chairman of CoP6, Jan Pronk, when interviewed after the proceedings, said that, 'Encouraging the widest participation in the process of promoting and co-operating in education, training and public awareness related to climate change is crucial. In developing country households women are often the primary providers and users of energy. Therefore, the participation of women and women's organisations is crucial.' (Wamukonya and Skutsch 2001, 13)

While this is undoubtedly true, there are many other aspects of climate change which might well have gender dimensions but which are not included in this statement. The fact that the gender dimension was evidently not a burning issue at the Hague meeting is perhaps all the more surprising given the fact that the spokespeople for three of the major NGOs – World Wildlife Fund, Friends of the Earth, and Climate Action Network – were women, and nearly 20 per cent of all the environment ministers present were female,[3] some of whom had key negotiating roles. Indeed, the success of earlier meetings, particularly the Kyoto meeting itself, is put down by some observers to the excellent networking done by female delegates committed to action on climate change (see Delia Villagrasa in this collection). Their lack of attention to gender issues may perhaps be attributed to their perceived need to focus on universal issues and not divert attention towards gender aspects, given the limited human resources for negotiation, and the crisis in which the whole debate on the Kyoto Protocol found itself at that time. In 1995, a Women's Climate Coalition called, rather wonderfully, 'Solidarity in the Greenhouse' was set up, and was pushing for special attention to women's energy needs.[4] But today their website is no longer active, and the group cannot be contacted by phone, fax, or e-mail. All sight of it has been lost by the UNFCCC in Bonn, and it was certainly not present at CoP6.[5]

However, the need for a gender analysis did come up as one of the very first conclusions at a preparatory meeting for WSSD 2002, which was held in Berlin shortly after the Hague CoP6 meeting (German Federal Ministry for the Environ-ment, Nature Conservation, and Nuclear Safety [BMU] and the Heinrich Böll Foundation 2001). Participants at this meeting called for the development of a gender analysis in all international energy-related processes, and, more immediately, for a Women and Climate Change Forum to be held at the resumed CoP6 in July 2001. However, just prior to this, President Bush announced the USA's decision to opt out of the Kyoto Protocol, pushing other concerns, including gender issues, to the background.

Despite women's caucus participation in the UN Commission for Sustainable Development process (CSD), they had limited influence in integrating decisive text relating to gender issues into the energy draft decision text deliberated by the ad hoc Open-Ended Intergovernmental Group of Experts on Energy and Sustainable Development in Feb 2001. The group was, however, able to persuade the G77 and China to introduce the issue of women and energy no fewer than five times in the Outcome Document at the CSD-9 meeting in New York in April 2001. Their persistent advocacy has thus borne some fruit. At the CoP7, held in November 2001 in Marrakech, a draft decision was reached (UNFCCC 2001) on improving the participation of women in the Parties' representatives.

The decision invites Parties to give active consideration to the nomination of women for elective posts in any body established under the Convention and the Protocol. In addition, the Secretariat is requested to maintain records on the gender composition of the various bodies.[6] Perhaps the election of a woman as the Co-ordinator of the African Negotiators Group, from the term starting after CoP7, will help to bring some gender issues into the mainstream of the climate negotiations of the CoPs in future, although whether there is a positive relationship here remains to be seen; past experience, as noted above, has not been very positive in this respect.

Potential areas of gender concern in the climate discourse

Apart from ensuring that there are more women on the various commissions within the climate change policy development process, gender considerations need to be included explicitly in future policy formulations and activities. Two rationales may motivate this: firstly, the idea that inclusion of gender considerations may increase the *efficiency* of the climate change process, and secondly, the concern that if gender considerations are not included, gender *equity* may be threatened, both of which are valid principles. There are three areas in the climate debate in which gender 'spectacles' might assist in promoting efficiency and equity, namely: responsibility for emissions, vulnerability to climate change, and participation in climate-change-related funded activities.

Responsibility for greenhouse gas emissions

Responsible as nations or as individuals?
Although the debate on what causes global warming may not yet be entirely resolved, the position taken here is that depicted in the IPCC reports and by the majority of scientists, which considers that human activities producing carbon dioxide and other greenhouse gases are responsible for a large share of the measured and predicted climate change. When discussing responsibility for the emission of greenhouse gases, however, one could raise the question of who, exactly, is responsible. At present this is being dealt with in the climate change negotiations with nations as the unit of consideration. Since the larger part of the greenhouse gases emitted into the atmosphere is the result of combustion of fossil fuels, and since the developed countries have large economies which use (and have in the past used) the lion's share of these fossil

fuels, most people hold that the developed countries should shoulder the burden of the problem, hence the allocation of emission reduction quotas to all developed countries.

A more radical idea is that every individual on earth should be given one and the same quota, and that through 'contraction and convergence' (Meyer 2000) we would eventually stabilise the levels of greenhouse gases in the atmosphere. This measure would imply much greater reductions in emissions in developed countries than are provided-for under the current agreements, while allowing developing countries to increase their emissions to a certain extent. Under this system every individual is, in the long run, equally responsible, but in the short term the problem has to be solved by those whose per capita emissions are highest.

The contraction and convergence idea still does not solve the problem of how the responsibility for *action* is to be sub-divided within any nation. To what extent can one group in a given economy be said to be more responsible for greenhouse gas emissions than another, or to be using more, or less, than their own individual quota? The only way in which it might be possible to administer a system by which all individuals, or groups of individuals, are in some way made directly accountable for their own greenhouse emissions would be via some kind of carbon tax on all products. In the context of gender issues, is it reasonable, or expedient, to argue that men and women may differ in this responsibility?

Gendered responsibility for primary emissions
On the one hand, it has been argued that major and global environmental threats stem primarily from industrial patterns of production and consumption. They are not due primarily to gender relations, nor will they be solved by improving gender relationships (Martine and Villareal 1997).

From this point of view, there is no need to take a gender position on 'responsibility' for climate change. One can contrast this with the ecofeminist school, which explicitly relates modern economies and their production processes to a male-dominated culture, arguing that economies based on 'feminine' – rather than 'masculine' – principles would look very different and would be much more environmentally friendly (Shiva 1989). Whether or not this is so, and whatever may change as regards the economy in the future, the fact is that we are currently stuck with the economic and industrial structures we have, with the problems that they entail, and with the need to clean up the mess they have produced.

The primary sources of greenhouse gases in the developed economies are the power industry, household energy use, and transport, followed by various industrial processes. Primary sources in the developing countries are the power industry, and land-use change, including clearing of forests. It would not be difficult to show that the power and the petroleum industries, and many industrial processes, are managed by men, both in the North and in the South. If a shareholders' survey is made, the probability is that the majority of their ownership will also be found to be male (in that more capital is in the hands of the male population in general). The question is, should men be considered more responsible than women for the problem? To answer this, we need to look at the services and products that these carbon-producing industries provide, and who uses them.

Gendered responsibility for use of products and services

There is some uncertainty surrounding the gender distribution of the services of these industries. Consider car ownership: although it has become increasingly less skewed over the last few decades in Europe and North America, and is slowly changing

in Eastern Europe, it is evident that cars are still used more by men than by women, with the side-effect that women are often disproportionately dependent on public transport (the situation in developing countries is even more extreme in this regard). So men – but of course not all men – are more responsible for greenhouse gas emissions produced by cars than women – or at least, some women – are. One could argue that the responsibility for emissions resulting from production of most manufactured goods must ultimately lie with the consumer, so that the question of responsibility depends on who the consumer is considered to be, making a gender analysis difficult. When it comes to other uses of energy, household energy-use in the developed countries is mostly related to heating and cooling, and is thus presumably consumed equally by men and women (although in most countries women spend more time at home than men). Basically, it is very difficult to make a strong case for a real gender difference, not least because income factors may have a much more important and confounding influence on energy use than gender.

The situation in developing countries is also difficult to assess clearly. Land clearance of forest for agriculture is traditionally a male activity, although much of the farm work that follows is carried out by women. Much of the benefit is for the household as a whole, even in cases where the cash crop profits accrue to men. To distinguish gender responsibilities becomes not just difficult but pointless. Besides, as they are the majority of household cooks, women could be blamed for greenhouse gas emissions from unsustainably managed fuelwood supplies! And who is responsible for the garbage problems in cities such as Nairobi, where the Dandora dump alone holds over 1.3 million cubic metres of garbage? Tonnes of methane emissions are produced from such dumps, and these cannot be allocated particularly to men or to women. Perhaps one could blame the local government officials (mainly male!) who have failed to provide an adequate alternative for trapping the methane. There are dangers in using this kind of argument to attribute responsibility by gender.

Responsibility for the direct or indirect production of greenhouse gases is more or less proportional to financial shares in the economy. In that women have a smaller financial share in the economy, one could argue that they are proportionately less responsible. However, using this as a principle on which to levy funds to cover the cost of global warming is fraught with difficulty. Such a policy would not increase the efficiency with which the problem of global warming can be tackled, nor would it easily serve to bring about greater gender equity. In the long run, it is evident that the costs of control of greenhouse gas emissions will have to be paid by the consumers of all goods or services via some kind of taxation system which reflects the real environmental costs of the whole lifecycle of that particular good or service. Thus women, if consuming less, will pay less.

Vulnerability to climate change outcomes: determinants and variables

Denton (2000, and see also her article in this collection), makes a strong argument that women in the South are more vulnerable than men to the effects of climate change. Her point, briefly, is that women are in general poorer than men, and more dependent on the kinds of primary resources that are most threatened by climate change, both in agriculture and in fisheries. As 'climate refugees' they will also be disproportionately affected. Women bear the burden of caring for the sick, and because increased levels of sickness are expected to result from climate change, women will bear the costs of climate change disproportionately. There is no

doubt that these are valid points. We need to ask however, whether the particular vulnerability of women to the effects of climate change is due more to the fact that they are, on average, *poorer* than men, or more to the fact that they are *women*, with particular roles and responsibilities which are especially prone to the effects of climate change? Should we approach vulnerability from the point of view of *gender*, or more generally from the point of view of *poverty*?

The view taken here is that analyses of vulnerability should explicitly recognise poverty as the primary variable. There is ample evidence at global and local levels that it is the poor who will suffer most from loss of livelihood related to gradual climate change, and also from sudden disastrous climatic events (such as floods and droughts), as they have little scope for adaptation, resistance, and insurance. This would seem to over-ride most other considerations. Most of the gender-specific characteristics that make people vulnerable to climate change (heavy dependence on local natural resources, lack of alternative income possibilities, responsibility for care of the sick, and so on) are in fact characteristics of women in societies of extreme poverty. In better-off societies, the effects of climate change will have less gender differentiation. What is important, therefore, is to recognise that poverty is not gender-neutral, and to understand and highlight the particular gender aspects of climate change vulnerability of the poor. Such recognition will lead to more efficient programmes for dealing with the effects of climate change, but also to greater gender equity.

In practice it should not be difficult to follow this course once the principle is recognised. Poverty research in general is increasingly becoming sensitive to gender issues, and recognition of the feminisation of poverty is a central issue in many development programmes. Methodologies and frameworks for such analyses (such as the Harvard method) are now widely available in the development literature. What is important is that such methods are taken on board and used in any climate change vulnerability studies that are undertaken in the context of the climate convention. In order to ensure that this happens, there is an urgent need that this be explicitly mentioned in the internationally accepted texts, which define the contents of such studies.

Participation in funded activities to mitigate, and adapt to, the effects of climate change

Combating the climate change problem is becoming a multi-billion-dollar business with funds for all kinds of projects in the private and public sectors. The question here is whether women are likely to be able to take an equal share in this, and what needs to be done to ensure that they do. A reasonable aim might be for women to access funds for climate purposes which at the same time have beneficial gender effects, for example opening opportunities for women to acquire technology which would otherwise be out of their reach financially. This would be beneficial from the efficiency point of view – cleaner technology spread, thus more carbon reduction – as well as the equity point of view, in that there would be more technology for women. The funds under the climate umbrella fall into a variety of types, which need to be addressed separately, since the opportunities for this type of 'win-win' strategy vary.

First, a number of donors are providing funds for so-called 'climate studies', which include the National Communications that all countries are required to produce under the UNFCCC, and other reports which document both the emissions of greenhouse gases and the effects of climate change on local populations. These are essentially scientific papers and the funding is therefore essentially research funding. The scientific community, particularly in developing countries, is of course more male than female but this is a general gender issue and not one that can be tackled specifically for the case of climate change.

More important will be the funds for *mitigation*, for *adaptation*, and for *capacity building*.

Mitigation funds

In the climate change negotiations, it is foreseen that mitigation (that is to say, reduction of greenhouse gases in the atmosphere) will mainly occur not through reduction of production and economic growth, which many environmentalists see as essential, but through economic growth where new, cleaner technologies are substituted for the old. The countries that are held responsible internationally for reduction of emissions (Annex 1 countries – the developed countries) have, with the exception of the USA, accepted reduction quotas, and plan to achieve these reductions not only in their own economies but by a number of so called flexible mechanisms abroad. The mechanism that concerns co-operation with developing countries is the Clean Development Mechanism (CDM), under which carbon saved by the transfer of clean technology to a developing county can be deducted from the quota of the developed country, which sponsors at least part of the costs of this clean technology.

The kinds of technologies most likely to be involved are those with the lowest cost per tonne of carbon saved, and include energy conservation technology (for example, in power generation, transport and manufacturing, fuel switching, and substitution of fossil fuel equipment by renewable energy technology where this is economic – although solar PV technology cannot compete price-wise in the carbon stakes at present). Under CDM, the setting up of 'sinks' (carbon sequestration in the form of forests) is also allowed, but only for the case of 'afforestation' and 'reforestation', which in practice means putting up forests where there were none before. CDM projects have to demonstrate that they have 'development effects' before they can be certified, but the definition of 'development effects' will be locally determined by individual developing countries. There is no specification in the law that CDM projects have to have any particular gender consideration: this is an aspect of development that also has to be determined by the host country.

Despite the fact that projects are supposed to have a development effect as well as a carbon mitigation effect, the reality is that carbon mitigation will be uppermost in the minds of the sponsors, who will select the cheapest and 'most efficient' ways of reducing greenhouse gas emissions. The cheapest ways of saving carbon are large-scale projects in the power and manufacturing sectors, and forestry sink projects. Although women might be involved in any of these as employees, there is no specifically gendered benefit to be gained from them (though like everyone else, women will hopefully enjoy increased access to electricity, reduced power outages, etc.).

Much more interesting for women, and particularly for poor women, would be the development and dissemination of a range of technologies in the areas in which they use energy now. These areas have received very little attention as regards project finance in the past. They include household energy, agricultural and food-processing, forest management, and water-pumping in rural areas, and energy appliances and processing equipment in peri-urban areas. The problem is that while in theory the CDM offers a whole new opportunity to market renewable energy technology to women, in reality this may not be as attractive to carbon investors as large one-off

investments in industry, despite the additional financial bonus that is implied by the emission reduction. They are unlikely to decide that targeting women will result in greater efficiency in offsetting carbon. This is despite the fact that there are obvious equity reasons for wishing to promote technologies to women.

Adaptation funds

From the beginning, there have been claims from many Southern countries that what is needed, even more than reduced emissions, is assistance with adaptation to the inevitable damaging effects related to climate change (raised sea levels, changing run-off patterns, increased disease levels, more weather turbulence, and so on). Some developing countries have prepared National Communications, and it is expected that this will be the basis on which adaptation programmes will be developed and implemented. Funds are to be provided for adaptation projects by a small levy on all CDMs, and by two special funds under the UNFCCC. At present, these funds are very small in comparison with the scale of the problems to be solved, particularly due to the withdrawal of USA from the Protocol and thus from the CDM mechanism. Moreover, the parameters or criteria under which a project may be considered an adaptation project have not yet been defined.

Nevertheless, taking a long-term view, there may be other opportunities for project financing for climate adaptation, and it is likely that more of the developed countries will pledge contributions bi-laterally. There are various kinds of investments likely to be considered: civil engineering work to shore up dikes and seawalls, and projects in agriculture and in forestry to enable vulnerable populations to maintain their livelihoods despite rapidly-changing climatic conditions. These are areas in which women are deeply involved and where, if treated in a gender sensitive manner, there might be real benefits to be gained, both in efficiency and in equity terms. An example would be forest management. Local community forest management projects already exist in many countries (India, Nepal, Mali, Burkina Faso, Uganda, to name but a few) in which women play an active role. In some cases, women are able to earn considerable income from continued sustainable harvesting of forest products to supplement meagre agricultural earnings. Such projects could easily be promoted as climate adaptation projects, in the sense that they modify micro-climates and protect water catchments, at the same time as diversifying income opportunities and thus protecting livelihoods (Skutsch 2002). They are much more likely to be directly beneficial to rural women than the sinks allowed under CDM.

Capacity-building funds

The pool of women professionals in the fields of engineering, energy, and other technical areas at all levels is small. Few women own, or are involved in managing, large businesses. Lack of financial and management capacity has been the main cause for this imbalance. If women are to be able to tap climate-change-related finances at all, it is clear that capacity-building focused on their needs will be necessary, including the need to lobby for their own interests within the climate negotiations. Specifically within the context of technology transfer and the flexible mechanisms, capacity will be needed to identify, assess, access, and assimilate technologies as well as to implement them.

Capacity-building has been seen by the international community as essential to enabling implementation of the UNFCCC, and the Kyoto Protocol. Funding has been, and will continue to be, allocated in increasing amounts. The question is, to what extent women, and particularly women with low incomes, can benefit from this, and what steps need to be taken to ensure that they do?

The need to address the likely bias of CDMs towards large-scale industrial projects and large-scale sinks, which are of little direct interest to most poor women, has already been mentioned. This implies that one aspect of capacity-building should be to assist women's groups to lobby for a more 'women-friendly' CDM policy, at least in the long term. There is also a need for attention to women's specific needs and capabilities as regards adaptation projects. Cleaner technologies in the agricultural and water sectors should target women as far as possible, and this may require gender-sensitive training for those responsible. This could be justified both on efficiency and on equity grounds.

Conclusions

There are many gender issues related to the UNFCCC and the instruments therein. Some, however, seem to be more interesting from a strategic point of view than others. While there is little to be gained by looking at the responsibility for emissions on a gendered basis, there would be benefit in publicising the fact that mitigation activities under the CDM are unlikely to bring much benefit to women unless that policy is explicitly adopted, and measures are taken to counter the flow of investment funds to the cheapest, large-scale investments for carbon saving. The opportunities to 'hijack' climate funds to direct renewable energy technologies towards women's real needs, so long under-estimated or ignored, should not be lost, even if this requires insertion of special clauses in the texts, and special sub-funds to finance them. Special attention also needs to be paid to the opportunities in adaptation investment which, based on assessments of vulnerability to climate change, will allow populations to survive the inevitable changes in the climate that are to come. Since these will to a large extent involve land-use solutions in rural areas, there is a lot of scope for women to be involved in these, and therefore gender-sensitive approaches in their design and implementation are important. Capacity-building, both of women themselves and of those entrusted with the development of policy and projects, is therefore essential at all levels in the international climate change process. Perhaps it is time to suggest that gender issues are specifically mentioned in the next international climate change treaty.

Margaret M. Skutsch, Technology and Development Group, University of Twente, PO Box 217, 7500 AE, Enschede, Netherlands. E-mail: m.m.skutsch@tdg.utwente.nl

(as this is a means of trying to shift the blame for environmental degradation from the industrialised to the developing countries). Within industrialised countries, the Coalition argued, responsibilities must not be shifted to private households entirely, as this will only conceal the role of industrial production processes (Solidarity in the Greenhouse 2001).

5 I am grateful to Sharon Taylor of the Climate Change Secretariat for this information.

6 The newly established 20 member CDM Executive Board includes two women. The Technology Transfer Expert group has 15 members, which includes three women.

Notes

1 This paper draws on an earlier publication, Wamukonya and Skutsch (2002). 'Gender spectacles' is a reference to Caren Levi (1992). I am grateful to N. Wamukonya for comments on this new paper.

2 For example, *Climate Policy, Joint Implementation Quarterly*, and *Climate Change*.

3 Female ministers were representing Bangladesh, Bulgaria, Chile, Costa Rica, Egypt, El Salvador, the EU, France, Gambia, Guinea, Honduras, Iceland, Iran, Japan, Mexico, Norway, Surinam, Tunisia, Venezuela, and South Africa.

4 The platform of the coalition had been to promote women's participation in policy and expert levels of UN decision-making, to reject Joint Implementation and nuclear power as climate strategies; to ensure that women's needs were explicitly dealt with at CoP1; and to lobby for financial support for women's renewable energy networks. They further stated that environmental policy-makers should not instrumentalise women of the South by holding them responsible for population growth

References

Denton, F. (2000) 'Gender impact of climate change: a human security dimension', *Energia News* 3(3)

German Federal Ministry for the Environment, Nature Conservation, and Nuclear Safety (BMU) and the Heinrich Böll Foundation (2001) 'Gender Perspectives for Earth Summit 2001: Energy, Transport, Information for Decision Making', Report on the International Conference at Jagdschloss, G., Berlin, http://www.earthsummit2002.org/workshop (last checked by the author April 2002)

Levi, C. (1992) 'Gender and environment: the challenge of cross-cutting issues in development policy planning', *Environment and Urbanization* 4(1)

Martine, G. and M. Villareal (1997) *Gender and Sustainability: Re-assessing Linkages and Issues*, Rome: FAO

Meyer, A. (2000) 'The Kyoto Protocol and the emergence of "Contraction and Convergence" as a framework for international political solutions to greenhouse gas emissions abatement', in O. Hohmeyer and K. Rennings (eds.), *Man-Made Climate Change: Economic*

Aspects and Policy Options, Mannheim: Zentrum für Europaischer Wirtsschafts-forschung

Shiva, V. (1989) *Staying Alive: Women, Ecology and Development*, London: Zed Books

Skutsch, M.M. (2002) 'Access to finance for community forest management under Kyoto and the UNFCCC', *European Tropical Forestry Research Network Newsletter* 35

UNFCCC (2001) 'The Marrakech Accord and the Marrakech Declaration', New York: UNFCCC

Solidarity in the Greenhouse (2001) http://www.alternatives.com/library/env/envclime/wa060015.txt (last found operating January 2001)

Wamukonya, N. and M. Skutsch (2001) 'COP6: the gender issue forgotten?', *Energia News* 4(1): March 2001

Wamukonya, N. and M. Skutsch (2002) 'Is there a gender angle in the climate negotiations?', *Energy and Environment*, forthcoming

Kyoto Protocol negotiations:
reflections on the role of women

Delia Villagrasa

At a first glance, the links between the results of the UN negotiations on climate change and gender issues may not be obvious. However, I believe that gender did indeed play a role in these discussions. This was not a role of the first order, but it was nonetheless a significant one. I would like to explain this impression by analysing briefly the three 'communities' which have shaped the United Nations Framework Convention on Climate Change (UNFCCC) Kyoto Protocol. The first community consists of the different countries' governmental delegations, the most important players in the negotiations. The second community consists of business and its representatives, and the third consists of environmental NGOs.

Women's participation in the UN negotiations on climate change is important. The reason for this is simple: the negative effects of climate change will affect women over-proportionally, and be felt more strongly in the South. The health effects (increased malaria and dengue cases, for example) of climate change, and therefore the caring for the ill, will fall mainly to women. The expected additional nutritional problems, and food and water security issues, will affect women more than men, as they are the main carers in these areas, particularly in regions where these are already critical issues (Vital Climate Graphics 2000). In many countries, women are the main household energy providers, often having to devote a substantial part of their working time to this task. The importance of women's inclusion in the negotiations on an issue affecting them heavily should therefore be obvious. However this is not the case, particularly for women from developing countries.

Since 1990, I have worked in the environmental NGO and business communities,[1]

and can report first-hand from those. Close observation of the governmental sector has given me a privileged insight into its working during the climate negotiations.

The negotiations leading up to the Kyoto Protocol were extremely intense. In 1997, official negotiations lasting over two months took place, mainly in Geneva, before the last two-week-long effort in Japan. Additionally, there were many informal workshops and consultations. In the lead-up to Kyoto there was a remarkable consistency in the participants at the negotiations. There were few changes in personnel at the decision-making level between 1990 and 1997. These circumstances led to a situation where the participants knew each other very well, and where decisions were often made under the influence of strong personalities, or based on trust and respect, and not solely based on the political and economic 'might' of a negotiating country (though the latter was obviously a major factor).

This consistency of personnel pertains to governmental as well as NGO communities, so that the sense of there being a 'climate family' has been created. This proximity has favoured close networking and communication between the actors involved. I believe that the unique framework and atmosphere of the climate negotiations have allowed the gender issue to become important. My thesis is that women were able to play a strong and generally positive role for climate protection based on their networking and interpersonal skills, and their ability to think and plan for the long term, even though they were generally under-represented in the decision-making positions of their respective communities.

The government sector: how women can influence negotiations

Each country participating in the climate change negotiations sends a delegation consisting of between one (especially in the case of cash-poor developing countries) and over 100 (in the case of the USA) negotiators, scientists, lawyers, and other specialists. The numbers vary according to the perceived importance of the negotiating session and the financial means of the country in question. Countries formed negotiating blocks, with the European Union (EU), the G77 (developing countries and China), and JUSSCANNZ (then Japan, the USA, Switzerland, Canada, Norway, and New Zealand) being the most important ones (Newell 2000).

Women clearly were, and still are, under-represented as formal delegation heads. However, it was some of the women in the delegations who played the most important roles in shaping the Kyoto Protocol. For example, the German and Swiss negotiation leaders, both women, through their true commitment to the cause of climate protection, tireless work, and networking skills, were able to influence the negotiations positively. Both acted in ways which differentiated them from their male colleagues in a crucial manner: they actively and often went out of their 'bunker', interacting strongly with other

delegations beyond the formal sessions. In particular, they were proactive in linking with delegations from developing countries, who were often greatly outnumbered, and furthering their integration into the decision-making processes.

The same holds true for the then Dutch Environment Minister, also a woman, who actively networked with developing country ministers. For many Southern delegations, it was impossible to follow the negotiations properly, as sessions often took place in parallel, making it impossible for a one- or two-person delegation to take part in every session. Furthermore, G77 is very diverse, representing interests from AOSIS to OPEC,[2] and is often bitterly divided on issues. While the German and Swiss leaders obviously had to represent their countries' interests in the negotiations, their personal integrity and openness earned the respect and trust necessary to 'build bridges' between nations. In the G77, there were women who played a strong unifying and progressive rule, from Zimbabwe and Peru particularly. The Philippine leader was notable for not bowing to extreme pressure from the USA on several key issues. This type of female interaction helped to forge links and mutual understanding, and this allowed the EU and G77 to build the alliance necessary to achieve the adoption of the Kyoto Protocol in face of the opposition of most of the JUSSCANNZ countries.

A lack of women's voices in the business sector

A lack of women participants was most obvious in the business sector, which was also geographically the least representative sector of the participants. The business sector was an almost exclusively male 'club', especially in the case of the decision-making lobbyists. Furthermore, the vast majority of business representatives were from the USA, with less than five per cent from developing countries. The businesses present were mainly representatives of the economic sectors afraid to 'lose' should climate protection become a reality, and therefore, with the exception of e[5] (The European Business Council for a Sustainable Energy Future) and the US Business Council for Sustainable Energy, they lobbied consistently and strongly against the Kyoto Protocol. Although, fortunately, it did not succeed in preventing the Protocol, the business lobby was strong enough to weaken the negotiating results considerably. Their lack of subtlety in negotiating (with one or two notable exceptions), their strident argumentation, and often aggressive style, were detrimental to their lobbying. Apart from their use of the OPEC countries to further their own agenda, they failed to network with the G77. I believe this was a major cause of their failure, and was probably partially due to a male lack of understanding of how to build networks with people from different backgrounds, cultures, and interests. e[5] has in the meantime been working proactively on issues such as the flexibility mechanisms of the Kyoto Protocol, helping to create a 'positive list' of projects to be favoured for sustainable development (e[5] 2000), and working with accession countries from Central and Eastern Europe towards the development of a solid emissions trading system.

The Climate Action Network: co-operating for change in the environment sector

Of the three sectors described here, the environmental NGOs were the most united. As a group, they were fighting for the AOSIS target, a 20 per cent carbon dioxide reduction by 2005, based on 1990 levels. The majority of the female participants formed part of this sector, though men predominated in some of the larger NGOs. However, the environmental NGOs had organised themselves into a unique, global structure: the Climate Action Network (CAN). CAN is an informal coalition of environmental NGOs, co-ordinated by

regional (mainly continental) 'nodes'. The most active and member-rich of those were Climate Network Europe, USCAN, CAN Canada, and CAN UK, later CANCEE, all led by women. CAN Africa was also led by a woman.

The NGO community recognised early on that due to their scarce resources, close co-operation was essential. Entrusting the team-building mainly to women, NGOs were able to create a cohesive, highly active force, which, despite being vastly out-numbered by the business community, was able to keep up a consistent flow of information to the public (through the media) and to government delegates, ensuring that Kyoto retained a high media profile, and forcing global decision-makers to bring the negotiations to a successful outcome.

It was women who ensured that NGOs worked together, despite their differences on some of the issues, and who ensured that debates did not get bogged down in detail and that coherent and strong messages went out to the world. For example, the female head of USCAN gave an emotive speech during the high-level segment of the negotiation, galvanising widespread media coverage. This type of co-operation, which also happens in-between negotiations, was essential in achieving the Kyoto Protocol.

I am not aware of another environmental issue where such a close co-operation on a global scale occurs, or where women have achieved such prominence in the NGO world. I do not believe that it is a coincidence that the issue where women have the most power in the NGO sector has also become one of the most actively debated ones in the public arena. It is also notable that this issue, which is not easy to explain to the public, whose consequences mainly lie in the future, whose causes are multifold, and where a cause-effect link is invisible, has attracted so many women. Many women seem to thrive on complexity and interlinked issues, whereas it has been my observation that men have rather been attracted to the straightforward 'battles', which may be easier to win.

Finding ways forward

Though women are playing an important role in the negotiations, these have now become much more complex and detailed, making it difficult to attract newcomers to the process. Newcomers have to invest a lot of time before they can be fully up-to-speed within the negotiations, and this is very difficult for organisations other than large businesses or wealthy governments to finance. Several possibilities exist for increasing women's participation in the negotiations. Negotiations are an excellent opportunity to meet other like-minded women and men, and to gather an enormous amount of climate-related information. Interested women could 'piggy-back' with a CAN regional organisation to get accreditation and information before a negotiation session starts. Sometimes there is even funding available for travel, though this is rare. Another way to enhance participation in the climate negotiations, both quantitatively and qualitatively, would be to introduce a mentoring system. This could take various forms, such as providing question and answer sessions for newcomers during the climate negotiations, or enabling newcomers to shadow experienced negotiators. I am aware that these recommendations aren't gender-specific, but I believe they could appeal particularly to women, who tend to share knowledge more easily than men.

At present, it is particularly necessary to create continued public pressure to advance the climate issue. At this stage, therefore, a very useful role for women, particularly in the South, would be in mobilising their governments towards proactive climate protection. For instance, we need to ensure that Clean Development Mechanism (CDM) projects really will contribute towards sustainable development through, for example, providing clean energy solutions for the two billion people who still do not have adequate access to energy. It would be helpful if women's development groups in the South could be alerted by the NGOs now active on climate issues, and a network created to promote the projects that are really needed to improve livelihoods – as communicated by women's networks in the South – as well as climate-relevant for the CDM. However, funding this kind of activity is problematic since it remains difficult to obtain funds towards capacity-building and networking on climate change. In the North, women's groups have so far barely been involved in the climate issue or the negotiations. If mobilised, they could put tremendous pressure on wealthier countries to provide the policies and measures needed for climate protection, such as a stronger development of renewable energy technologies.

Certainly, and on a less resource-intensive level, an easy way for women and women's organisations to become more involved in the issue and the negotiations would be to learn from the success of CAN. They could join the existing e-mail networks to learn about key issues, and where access to e-mail is not available, regular newsletters from many of the CAN member organisations could be distributed.

Conclusions

I believe that we have no reason to be complacent. Women's networking strength needs to be harnessed even more strongly, within the political, business, and environmental NGO sectors globally, with the aim of preventing climate change as far as possible, and adapting to it where necessary. Nonetheless, we have existing capacity upon which to build, and young women are joining from the environmental as well as the scientific, development, and legal communities to win the fight. I am confident that women will continue to contribute significantly to ensuring that the climate negotiations will be translated into real action on the ground.

Delia Villagrasa is Executive Director of e5 (European Business Council for a Sustainable Energy Future). Contact: 64, Boulevard de la Cambre, B-1000 Brussels. Tel: +32 2 644 2888; Fax: +32 5 687 4641;
E-mail: deliavilla@gn.apc.org

Notes

1 From 1990-4 for WWF International; from 1994-5 for FUNDES, a development organisation; from 1995-9 for Climate Network Europe; and from 2000-02 for e[5].

2 AOSIS is the Alliance of Small Island States, a group of states in danger of vanishing as a result of climate change and rising sea levels, which advocates strong greenhouse gas reduction targets. The Organisation of Petroleum Exporting Countries (OPEC) forms a determined opposition to climate protection measures.

3 The European Business Council for a Sustainable Energy Future was the first business association to lobby consistently in favour of the Kyoto Protocol as well as for strong climate protection measures in Europe and globally.

References

e[5] (European Business Council for a Sustainable Energy Future) (2000) 'COP6 Position Statement, November 2000', Brussels: e[5]

Newell, Peter (2000) *Climate for Change, Non-State Actors and the Global Politics of the Greenhouse*, Cambridge: Cambridge University Press

Vital Climate Graphics (2000) *The Impacts of Climate Change*, UNEP/GRID-Arendal, United Nations Framework Convention on Climate Change (UNFCCC), http://www.unfccc.int (last checked by author November 2000)

Gender and climate hazards in Bangladesh

Terry Cannon

Bangladesh has recently experienced a number of high-profile disasters, including devastating cyclones and annual floods. Poverty is both a cause of vulnerability, and a consequence of hazard impacts. Evidence that the impacts of disasters are worse for women is inconclusive or variable. However, since being female is strongly linked to being poor, unless poverty is reduced, the increase in disasters and extreme climate events linked with climate change is likely to affect women more than men. In addition, there are some specific gender attributes which increase women's vulnerability in some respects. These gendered vulnerabilities may, however, be reduced by social changes.

To many from outside, Bangladesh is almost synonymous with disasters. In a country smaller than Britain, and with more than twice as many people, around one-third of the land is flooded every summer. The monsoon rains cover the low-lying land, and swell the three major river systems that struggle to find outlets to the sea. In some years, such as in 1998, nearly half of the land area of Bangladesh is under water. Tropical cyclones strike the coast at least once a year, bringing rainwater floods, salt-water incursions, and wind damage. Since the 1991 cyclone disaster, effective warning systems, coupled with the use of many more cyclone shelters, have reduced the toll to a fraction of earlier tragedies, and now the number of deaths each year is usually less than a thousand.

The inland rain and riverine floods have attracted considerable foreign attention and aid, as evidenced by the Flood Action Plan (FAP) of the early 1990s. Yet, paradoxically, the deaths caused by these events rarely exceed a few thousand – in contrast to the death toll of cyclones – and never reach tens of thousands. Floods are very visible and may appear to be a disaster, even though they are vital to the livelihoods of almost all of the rural population. Therein lies a second paradox: most of the rural population actually considers it a disaster when there is no flood. Without the annual cycle of inundation and silt, the fertility of fields is diminished, and they produce a much lower yield as a result of lack of water. Moreover, fish breeding is disrupted and output diminished when flooding does not create ponds and interconnections between waterways. This is a severe disadvantage to the poor who depend on fish as their main source of protein (and sometimes income).

This does not mean that floods should always be regarded as beneficial, or that people do not lose lives, assets, or become even poorer as a result of them (for example, those who lose land from erosion by the shifting of river channels in floods). However, while a flood can produce an obvious deepening of poverty, its absence has invisible consequences that may be just as bad. A distinction is made in Bengali between 'good' and 'bad' floods to reflect the difference. In general, the majority of the rural population would lose out rather than benefit from the prevention of flooding, by engineering measures such as embankments and river containment, envisaged in the FAP. The benefits of 'good' floods outweigh the disadvantages of the 'bad' (Blaikie *et al.* 1994). In a rare sample survey of rural people's attitudes to floods, 86 per cent of households were satisfied with the way that they adjusted to normal inundation, and did not want any change to that situation (Leaf 1997).

Climate change, hazards, and their gender dimensions

The principal climatic hazards affecting Bangladesh – floods and cyclones – are likely to increase in frequency, intensity, duration, and extent. The summer monsoon rainfall is projected to increase, swelling the main river systems in the wider catchment, and boosting the rainfall impact within the country. More rapid glacial melting in the Himalayan headwaters will also increase spring and early summer flows, further increasing the flood risk. In winter, problems of drought will increase. The current winter dry season (which already limits agriculture and particularly affects poorer farmers who cannot afford to irrigate) is likely to become significantly worse (World Bank 2000). Cyclones are low-pressure systems, which means that as well as causing rainfall flooding and wind damage, they raise sea levels and bring storm surges that flood the coast with salt-water. With rising sea levels, it is estimated that within a century the coastline will retreat by, on average, about ten kilometres, causing the loss of 18 per cent of the country's land area. This will mean that the impacts of cyclones will be felt further inland than they have been to date (*op. cit.*, ii).

How these increased hazard impacts will affect women in particular, is extremely difficult to predict. The link between poverty and vulnerability is clearly crucial, and affects women disproportionately. If there is no serious progress in reducing poverty, then it can be assumed that women will become increasingly affected by the impact of intensified hazards, in terms of their ability to resist and recover from them. This outcome may be modified if there are more general reductions in economic inequalities between men and women.

It is also important that non-economic ('cultural') factors which produce gender inequality are also addressed – for instance, so that women can adequately seek shelter without shame and harassment, and are not condemned to poverty and increased vulnerability when widowed or divorced. These are issues that are already on the sustainable development agenda, and so it could be argued that reducing women's vulnerability to hazards will follow from this agenda. However, such an approach does not adequately address the specific gender dimensions of disaster preparedness. Evidence from Schmuck (2002), German Red Cross (1999), Baden *et al.* (1994), Rashid and Michaud (2000), Enarson (2000), Enarson and Morrow (1997), and Khondker (1996), all suggest that there are specific gendered factors which it is essential to take into account in order to reduce the vulnerability of women.

Understanding disasters and vulnerability

Disasters happen only when a natural hazard impacts negatively on vulnerable people. The severity of a disaster is therefore a reflection both of the location and intensity of the hazard, and of the number of people of given levels and types of vulnerability. For instance, tropical storms of similar intensity affect the USA and Bangladesh, but with very different outcomes. In 1992, Hurricane Andrew struck Florida, and caused more than 28 billion pounds' worth of damage, but killed fewer than 20 people (Morrow 1997). The year before, the cyclone that struck the south-east coast of Bangladesh killed 140,000 people, and ruined the livelihoods of millions (German Red Cross 1999). This does not mean that the people of Florida were unscathed and that they did not suffer (physically and mentally) from loss of homes, schools, jobs, and possessions. But the illustration shows how the impact of an equivalent hazard on different communities is related to differing levels of social vulnerability. This vulnerability can be considered to have five components, which vary from higher to lower levels according to political and social factors affecting different groups of people: namely, the initial conditions of a person, the resilience of their livelihood, their opportunities for self-protection, and their access to social protection and social capital (Cannon 2000; Blaikie *et al.* 1994). These differ hugely between the contexts of Bangladesh and the USA.

To understand a disaster, we need to understand the components of vulnerability of different groups of people, and relate these to the hazard risk (Cannon 2000; Blaikie *et al.* 1994). Vulnerability differs according to the 'initial conditions' of a person – how well-fed they are, what their physical and mental health and mobility are, and their morale and capacity for self-reliance. It is also related to the resilience of their livelihood – how quickly and easily they can resume activities that will earn money or provide food and other basics. The hazard itself must be recognised, and the fact that vulnerability will be lower if people are able to put proper 'self-protection' in place – e.g. the right type of building to resist high winds, or a house site that is raised above flood levels. People also usually need some form of 'social protection' from hazards: forms of preparedness provided by institutions at levels above the household. These supplement what people cannot afford or are unable to do for

themselves, and provide opportunities to implement measures that can only be provided collectively (e.g. codes to improve building safety, warning systems).

Social protection depends on adequate government or non-government systems being in place, while self-protection generally relies on people having an adequate income, knowledge of the hazard, and propensity and capacity to take precautions. In many hazardous places, people's vulnerability is also reduced if they are able to draw on adequate social capital. People may need to rely on each other, on family, and on organisations, at all stages of a disaster – from search and rescue after impact, to coping and sharing in the recovery period. Social capital may not always be neutral and benign: there are examples of disaster recovery where some people identified in a particular social group received assistance not made available to others, as after the Gujerat earthquake of 2000 (Vidal 2002).

Gender inequality, women's status, and capacity for protection

How are these components of vulnerability affected by gender relations, and how different are the vulnerabilities of men and women in relation to disasters in Bangladesh? From an analysis of existing gendered vulnerabilities, can we project what may happen in terms of climate change and the possible increase in frequency and intensity of climate hazards? Vulnerability in Bangladesh correlates strongly with poverty, and it is widely accepted that women make up a disproportionate share of poor people. How much of women's vulnerability to hazards can be apportioned to them being poor, and how much is due to specific 'gendered' characteristics of self protection, social protection, and livelihood resilience? And how will this be affected by climate change?

In fact, it is difficult to separate these two aspects of female vulnerability, precisely because gender plays a significant role in determining poverty. A recent Asian Development Bank report suggested that over 95 per cent of female-headed households are below the poverty line. The proportion of female-headed households in Bangladesh was officially reported as ten per cent, but other evidence cited suggests that a more realistic figure is 20–30 per cent (Asian Development Bank 2001). Many of these households consist of women who have been divorced or widowed, and who are culturally discouraged from remarrying. Ninety per cent of those who are single as the result of bereavement or divorce are women (*ibid.*). As a result, vulnerability to hazards involves a complex interaction between poverty and gender relations, in which women are likely to experience higher levels of vulnerability than men.

Women's nutritional status and coping capacity

Women's poorer nutritional status is a key aspect of their reduced capacity to cope with the effects of a hazard. In Bangladesh, women of all ages are more calorie-deficient than men, and the prevalence of chronic energy deficiency among women is the highest in the world (del Ninno *et al.* 2001). Although this study of the 1998 flood found no evidence of any increase in discrimination against females, it is clear that the situation is potentially disastrous. 'Given the already precarious nutritional state of large numbers of girls and women in Bangladesh... any further increase in discrimination against females in food consumption would have serious consequences.' (*op. cit.*, 64). Women also receive less and poorer-quality healthcare in comparison with men. Bangladesh is one of the few countries in the world where men live longer than women, and where the male population outnumbers the female (Asian Development Bank 2001).

Women's domestic burden and increased hardship

There is evidence that floods increase women's domestic burden. The loss of utensils and other household essentials is a great hardship, and floods also undermine women's well-being in general because of their dependence on economic activities linked to the home (Khondker 1996). In their study of gender in Bangladesh, Baden *et al.* found that women are likely to be less successful, and find it more difficult to restore their livelihoods, after a flood. Losses of harvest and livestock have a disproportionate impact on women, many of whom rely on food processing, cattle, and chickens for their cash income (Baden *et al.* 1994). Fetching water becomes much more difficult, and it may be contaminated. Water-borne illness might be expected to be more widespread among women, who are nutritionally disadvantaged. Women are likely to suffer increased mental strain, and bear the brunt of certain social constraints, for instance they are shamed by using public latrines, or being seen by men when in wet clothing (Rashid and Michaud 2000).

Women's reduced ability to provide self-protection

Poverty is a key factor affecting people's ability to provide adequate self-protection, and it is likely that in female-headed households, the ability of women to create safe conditions in the face of impending floods or cyclones is reduced. The quality of housing, a location on raised ground, adequate storage for food – all are crucial to self protection, but are more difficult for poor women to achieve. Both self- and social protection are also affected by gender issues related to 'culture'. During cyclones, women are handicapped by fear of the shame attached to leaving the house and moving in public. It may be too late when they eventually seek refuge. Societal attitudes restricting interaction between men and women make women more reluctant to congregate in the public cyclone shelters (raised concrete structures that protect from wind and flood) where they are forced to interact with other men. However, NGO activities to increase understanding and make warnings more effective seem to have improved this over the past ten years (German Red Cross 1999). Women's mobility is restricted as a result of their responsibility for their children. Their clothing restricts their mobility in floods, and in addition, women are less likely than men to know how to swim. It is estimated that 90 per cent of the victims of the 1991 cyclone disasters were women and children (Schmuck 2002).

Social change: a glimmer of hope?

There is evidence that some aspects of social change in Bangladesh are improving women's lives and reducing gender inequalities. The average number of children that a woman bears has declined significantly over the last 20 years, from 6.34 in 1975 to around 3.3 in 2001 (BBC 2001). This has significantly reduced women's child-care burden. It has also made their lives safer: more women die as a result of childbirth in Bangladesh than anywhere else in the world. Whether the significant cultural shifts inherent in this decline in fertility rates can have any impact in other areas of society, including on gender differences in vulnerability to climate hazards, is impossible to predict. If progress continues to be made in improving women's lives and reducing gender inequalities, through other initiatives such as micro-credit schemes for women, and associated empowerment activities by NGOs, then there is potential to reduce women's unequal vulnerability as the hazards increase with climate change.

Terry Cannon, Natural Resources Institute, University of Greenwich, Central Avenue, Chatham, Kent ME4 4TB.
E-mail: t.g.cannon@greenwich.ac.uk

References

Asian Development Bank (2001) 'Country Briefing Paper: Women in Bangladesh', Manila: Asian Development Bank

Baden, S., C. Green, A.M. Goetz, and M. Guhathakurta (1994) 'Background report on gender issues in Bangladesh', *BRIDGE Report* 26, Sussex: Institute of Development Studies

BBC (2001) 'Family planning in Bangladesh', World Service, 1 October, http://www. BBC.co.uk/worldservice/sci_tech/ highlights/011001_bangladesh.shtml (last checked by the author March 2002)

Blaikie, P., T. Cannon, I. Davis, and B. Wisner (1994) *At Risk: Natural Hazards, People's Vulnerability and Disasters*, London: Routledge

Cannon, T. (2000) 'Vulnerability analysis and disasters', in D.J. Parker (ed.), *Floods*, London: Routledge

del Ninno, C., P.A. Dorosh, L.C. Smith, and D.K. Roy (2001) 'The 1998 floods in Bangladesh: disaster impacts, household coping strategies, and response', *Research Report* 122, Washington DC: International Food Policy Research Institute

Enarson, E. (2000) 'Gender and natural disasters', *Working Paper* 1 (Recovery and Reconstruction Department), Geneva: ILO

Enarson, E. and B.H. Morrow (1997) 'A gendered perspective: the voices of women', in W.G. Peacock *et al.* (eds.), *Hurricane Andrew: Ethnicity, Gender and the Sociology of Disasters*, London: Routledge

German Red Cross (1999) 'Living with cyclones: disaster preparedness in India and Bangladesh', Bonn: German Red Cross

Khondker, H.H. (1996) 'Women and floods in Bangladesh', *International Journal of Mass Emergencies and Disasters*, 14(3): 281-92

Leaf, M. (1997) 'Local control versus technocracy: the Bangladesh Flood Response Study', *Journal of International Affairs* 51(1): 179-200

Morrow, B.H. (1997) 'Disaster in the first person', in W.G. Peacock *et al.* (eds.), *Hurricane Andrew: Ethnicity, Gender and the Sociology of Disasters*, London: Routledge

Rashid, S.F. and S. Michaud (2000) 'Female adolescents and their sexuality: notions of honour, shame, purity and pollution during the floods', *Disasters* 24(1): 54-70

Schmuck, H. (2002) 'Empowering women in Bangladesh', http://www.reliefweb.int (last checked by the author March 2002)

Vidal, J. (2002) 'Helping hands', *The Guardian*, 30 January 2002

World Bank (2000) 'Bangladesh: climate change and sustainable development', Report no. 21104 BD, Dhaka: South Asia Rural Development Team

Uncertain predictions, invisible impacts, and the need to mainstream gender in climate change adaptations[1]

Valerie Nelson, Kate Meadows, Terry Cannon, John Morton, and Adrienne Martin

Vulnerability to environmental degradation and natural hazards is articulated along social, poverty, and gender lines. Just as gender is not sufficiently mainstreamed in many areas of development policy and practice, so the potential impacts of climate change on gender relations have not been studied, and remain invisible. In this article we outline climate change predictions, and explore the effects of long-term climate change on agriculture, ecological systems, and gender relations, since these could be significant. We identify predicted changes in natural hazard frequency and intensity as a result of climate change, and explore the gendered effects of natural hazards. We highlight the urgent need to integrate gender analyses into public policy-making, and in adaptation responses to climate change.

Although 'gender' has been recognised as an important factor within development policy since at least the 1970s, there is still a lack of practice to match the rhetoric. Thus, it is not surprising that the gender dimensions of climate change have largely been neglected. This is despite the fact that the effects of climate change are *very likely* to be gendered. It is possible to infer this because of the strong relationship between poverty and vulnerability to environmental change, and the stark fact that women as a group are poorer and less powerful than men.

In this article we will explore some of the reasons why the impacts of climate change on gender relations have not been fully articulated. We consider climate change predictions. Long-term climate change will have an impact on agriculture, and ecological and human systems, and is therefore likely to have ramifications for gender relations. Over longer timescales, broader issues of international policy, regional context, and politics will also affect social change. Extreme weather events are also expected to increase in intensity. Some studies of the gendered impacts of natural hazards are available, providing clues to the possible outcomes of climate change. Given the uncertainty about how climate change will manifest itself in different regions, prediction and quantification of consequent social changes will be difficult. Attribution of causality is also difficult because over a similar timescale, major shifts in gender relations have occurred in some countries. We finish this paper with a discussion of gender-aware public policy-making on climate change responses, and a call for further context-specific research.

Neglect of the impacts of climate change on gender relations

The impact of climate change on gender relations has been neglected, due to the 'gender-blindness' still afflicting much development policy-making, and the slow response by development agencies to the development challenge presented by climate change. This is partly due to the uncertainty of climate change prediction, especially at the regional level (although this is improving) (Dalfelt 1998), and the lack of mainstreaming of environmental issues into development thinking. Further, many climate change studies focus only on very broad-brush areas of environmental impact.

Environmental degradation can increase both women's workload and their vulnerability, as their access to already scarce resources decreases. Poor or missing gender analysis can mean that planners depend on women assuming a central role in coping strategies, without taking into account the increased burden that this imposes on women. Assumptions may be made that women are 'closer to nature' than men, and therefore that the responsibility for environmental protection is exclusively, or largely, that of women. Relief and development projects may also rely too heavily on women's unpaid labour, when it is assumed that women are *naturally* predisposed to serve their families or communities by protecting the environment on which they depend for livelihoods.

Climate change predictions

Temperature increases are already affecting biophysical systems (IPCC 2001, 3-4). Certain natural systems are particularly vulnerable (mangrove forests, small island states, coastal areas, and so on). Negative outcomes may include declining crop yields in many tropical and sub-tropical areas, decreasing water availability in arid regions of the sub-tropics, and an increase in vector- and water-borne diseases, heat stress mortality, flooding, and wildfire incidence. The spread of disease, and reduced drinking water availability in some areas, could have gender-differentiated impacts where women have less access to medical care than men. Positive outcomes are also possible – in some regions currently lacking water (e.g. parts of South-East Asia), more water may become available (IPCC 2001).

Climate change consists of both 'short and medium term climate variability together with long-term gradual climate change (changes in annual average temperature)' (Dalfelt 1998, 2). The increasing frequency of extreme weather worldwide (especially droughts and floods) seems to be correlated with the El Niño Southern Oscillation (ENSO), although further research is required to substantiate this connection.

Gender impacts of climate change on agriculture and ecological systems

Long-term gradual climate change will affect agricultural and ecological systems. It may be difficult to disentangle the effects of increasing natural hazards, local environmental degradation, and long-term climate change. Nonetheless, it is clear that there will be a complex patchwork of alterations, difficult to predict accurately, which will challenge people's ability to cope, and governments' capacity to adapt. Crop and livestock responses will vary according to species, cultivar, pests, and so on. A whole range of adaptations in cultivation and husbandry are possible (IPCC 2001). Natural hazards are likely to have a more visible impact on people than the slower onset of changes in temperature and rainfall regimes, although the latter may be more significant in the long run for those dependent on farming.

Women in developing countries, who are often primary natural resource users and managers (for example, collecting firewood, forest products, and water), are often disproportionately affected by environmental degradation. Households dependent on women's labour in subsistence or cash cropping or on plantations are also badly affected by storms and droughts.

Some less-productive tropical climates will become unsuitable for agriculture as a result of climate change (Mendelsohn and Dinar 1999). Agricultural GNP (gross national product) may not be significantly damaged overall (because there are areas of productive temperate farmland in many developing countries, even in equatorial zones), but small-scale farmers may lack the capital and resources necessary to adapt to climate change, especially in comparison with larger enterprises (Mendelsohn and Dinar 1999). Such potential scenarios, and responses to them, need to be analysed with a 'gender lens' to try to identify the possible negative and positive outcomes in gender-differentiated terms.

Drought and gender

Drought can be considered to be a slow-onset disaster. The frequency and intensity of drought in great swathes of dryland Africa and West, Central, and South Asia, has increased over recent times and is predicted to increase further with climate change. In Iran, Afghanistan, and parts of Pakistan and India, there has been severe drought for four successive years. In Morocco, ten of the 16 years from 1984-2000 were considered drought years (MADREF/World Bank 2000). Northern Kenya experienced periods of severe drought in 1983-4, 1991-2, 1996-7 (Hendy 2001), and 1999-2001.

Beyond the increasing incidence of drought, dryland populations are increasingly vulnerable to drought resulting from socio-economic trends and local environmental pressures. There are circular relationships between drought and desertification (Dregne 2000), and while the relationships between drought and human vulnerability are complex, there is evidence that the impacts of drought are gendered.

These impacts will be locally specific, but in different parts of the world, and for different socio-economic strata, they could include the following:

- male out-migration, generating increased work for women on farms – though the effects of this on women's autonomy can be complex, and female out-migration also occurs;
- cropping changes, with effects on gender division of labour and possibly income;
- livestock production changes (large-stock to small-stock, open-grazing to pen-feeding), with effects on gender division of labour and possibly income;
- increased difficulty in accessing resources (in particular fuelwood and water), hence increased workload for women;
- increased conflict over natural resources, exacerbated in some places (e.g. East Africa) by the ready availability of firearms;
- health impacts: direct impacts on women's health, and increased work for women as carers.

Besides the direct effects of drought and desertification, government and donor action in drought situations is now often massive in scale. Despite attempts to design cost-effective mitigation measures to support peoples' livelihoods at key points in the drought cycle (for pastoral livelihoods, see Morton 2001a), the backbone of international drought management is formed by direct food aid and labour-intensive public works projects. In recent years, drought has rarely become international news, and hence the vast scale of drought relief operations is not always appreciated. It would be surprising if such massive interventions were not themselves having long-term, gendered effects on livelihoods, mediated by social, cultural, and institutional factors.

In Morocco, the Government allocated (from its own resources) over US$400 million for labour-intensive public works to mitigate the effects of drought, in the 15 months from April 2000-June 2001 (a not exceptional level in recent times) (Morton 2001b). The work opportunities created in drought-stricken areas in Morocco (e.g. manual labour on roads) are regarded as being almost exclusively for males. Project planners are relatively unconcerned that households without able-bodied males may not benefit directly from these opportunities, because they point to the extended kinship ties that already offer support to widows and to households that are effectively female-headed as a result of labour migration. The combination of massive expenditure, and the association of manual labour with men only could, however, lead to the reinforcement of existing gender inequalities.

Food relief in drought situations can fuse with more direct effects of drought and environmental degradation to create a process of sedentarisation of pastoralists, with complex effects on gender relations. In some areas of northern Kenya, food-for-work has become an important part of livelihoods, and has led to the settlement of previously nomadic pastoralists around missions or administrative centres (Baxter 1993). This has acted as a trigger for longer-term settlement processes. Differences are emerging amongst settled, semi-settled, and nomadic pastoralist women. While women in more mobile pastoralist groups consider themselves 'economically' better off, as mobility increases their access to pasture and fuelwood, they are in their own words 'treated as children', with little input into pastoral movement decisions or in donor initiatives. Semi-settled women also have low input into major household decisions, while reduced mobility increases

the difficulties of water and firewood collection. Settled women suffer the greatest resource pressures, but they enjoy the greatest autonomy. While they have started to earn an income from market sales of firewood, they have to walk further for it; although they have considerable control over small-stock, movement to pasture is difficult. Some also sit on the Environmental Management Committee. 'These townswomen were somewhat disdainful of those on the periphery, and [said they had realised that] "two heads are better than one" in household decision-making. They considered this difference to be a product of "ignorance" among the non-settled.' (Meadows 1999)

Natural hazards and gendered impacts

How will the likely increased frequency and intensity of natural hazards (one of the outcomes predicted for climate change) affect poor people, and specifically women?

Men's and women's differing experiences of natural hazards are not well-researched, particularly in developing countries. However, it is well known that women experience high levels of *pre-disaster* poverty, often experiencing unequal status in the workforce, being more likely to be employed in the informal sector, and having 'less equitable access to land and other natural resources compared to men' (Enarson 2000). The impact of a natural hazard depends upon the social context within which it occurs. This socially-constructed vulnerability extends to the contextual gender and power relations (Blaikie *et al.* 1994; Enarson 2000). Those living in areas most at risk are often those with least social and economic power, and who are least able to cope with, and recover from, disasters. Women are often key to household survival when disasters strike, although their responsibilities in the domestic sphere make them economically vulnerable before such an event occurs. Groups of women likely to be particularly vulnerable to natural hazards include refugees, those on low incomes, homeless, elderly, and disabled women, recent migrants, and so on.

Women's work can be affected in a variety of ways by natural hazards. Productive assets may be lost, pushing women into low-wage labour. More women than men work in the informal sector and in small enterprises. These sectors are often worst-hit, and least able to recover as a result of disasters. Natural hazards cause women to lose jobs and work-time disproportionately, and conditions of work often deteriorate. On the other hand, some women – middle class women in particular – can benefit in terms of changed access to employment opportunities (Enarson 2000).

In some places and situations, women are more at risk because of culturally-specific pre-disaster gender norms. Female mortality was much higher than male mortality in the 1991 cyclone floods in Bangladesh. Of the flood-affected population in the 20-44 age group, 71 females per thousand died compared with 15 males per thousand (cited in Baden *et al.* 1994, 49). Most were drowned. Cultural norms relating to the preservation of female honour through seclusion mean that women may delay leaving the home to seek refuge, until it is too late. Norms relating to what may be considered appropriate activities for women and men mean that women are also less able to learn to swim. An increase in flood frequency and intensity might thus increase female mortality.

Gender norms also affect the behaviour of men during disasters. Ideas about masculinity may encourage risky 'heroic' action in a disaster, and may also mean that men are less likely to seek counselling afterwards (Enarson 2000). More men died than women in Hurricane Mitch, for example,

showing that relationships between natural hazards and gender do vary (Delaney and Shrader 2000).

Mortality and morbidity are only part of the range of impacts of hazards. Gender relations are unlikely to improve spontaneously as a result of increased hazard risks. There are gender dimensions to what happens in the aftermath of a hazard strike in the relief, coping, and recovery phases, where there is strong evidence of considerable inequality between men and women.

Gender and the aftermath of hurricane impacts: the example of Hurricane Mitch

Gender differences and inequalities are most pronounced in the *aftermath* of a hurricane, and these differences may persist for months and years. These include many aspects, ranging from the increased workload of women, to their greater exposure to violence as a result of raised aggression levels in men. The 1998 Hurricane Mitch directly affected more than two million people in Honduras and Nicaragua alone. Damage estimates were placed at nearly US$5 billion. Those most affected were the most marginalised (small producers, street children, and female-headed households) (Delaney and Shrader 2000, 5). Women endured a disproportionate amount of the burden immediately following the storm and in later rehabilitation, because of their triple roles in maintaining the household, engaging in community organising, and productive work in the informal economy. Women had the main responsibility of caring for children and the elderly. Men generally tried to return to their pre-disaster role of earning wages outside the home, whilst women found it difficult to return to waged work. This, combined with the fact that more men than women had died, led to large increases in female-headed households (rising to 40 per cent of total households in Nicaragua and half in Honduras) (*ibid.*).

Household food hierarchies exist (placing females below males), and disasters can reduce the overall amount of food available, exacerbating the unequal position of women. Women are likely to have poorer nutritional status and resistance to disease, and so are likely to be more at risk than men (Blaikie *et al.* 1994). Combined with their poorer access to medical care, the health of women is disproportionately affected. During rehabilitation, whilst women maintained the household and social networks, men were involved in dangerous reconstruction efforts; some men were also taking part in increased gambling, increased consumption of alcohol, and some were displaying greater aggression (*op. cit.*).

The social disruption and altered intergroup relations that can occur as a result of a disaster may enable women to challenge or override existing gender norms. This may affect traditional divisions of labour, or enable the organising of new forms of social capital and disaster preparedness (PAHO 2001, 2). Gender norms are challenged when women take on tasks traditionally ascribed to men, gaining new skills and changing prevalent views as to women's capabilities. This occurred after Hurricane Mitch when women were observed building shelters and wells (PAHO 2001, 2).

Towards gender-sensitive climate change policy responses

Responses to the impact of climate change on agriculture will need to be gender-aware, otherwise government policies and development programmes aimed at supporting adaptation by farmers could further exacerbate gender inequalities. Public policies need to be more responsive to the livelihood decisions faced by local people, and the potential impacts of these on power and gender relations.

Many factors will influence social change, including global and regional political economy. The impact of economic and agricultural policies need to be taken into account in terms of how they affect the future resilience of poor peoples' livelihoods to climate change. In the past, agricultural policies in developing countries have promoted cash- and mono-cropping, and an export orientation. However, it is possible that such agricultural systems are less resilient to climate change than more diverse agro-ecological systems. The latter include a wide band of species, enabling farmers to spread the risks of disease and crop failure under climatic stress. In Kenya, for example, colonial and post-colonial agricultural policies had exactly this kind of export-, cash-, and mono-cropping orientation, which undermined the intercropping of diverse bean species (e.g. *Lablab niger* and *Dolichos lablab*) by women, and reduced seed stock diversity (World Bank 2000). Women's key role in maintaining biodiversity, through conserving and domesticating wild edible plant seed, and in food crop breeding, is not sufficiently recognised in agricultural and economic policy-making; nor is the importance of biodiversity to sustainable rural livelihoods in the face of predicted climate changes.

Social change, climate change adaptation, and gender

Further research is required to explore how climate changes will manifest themselves in different regions and on different time-scales, and how social and natural systems will co-evolve. There are no 'given or *a priori* sets of driving forces (such as technologies, markets, policy imperatives, or cultural values) that generate particular social arrangements or patterns of change, only complex sets of connectivities between material, cognitive, social, and non-human elements' (Long 1997, 109). This is why predictions in patterns of social change and future social arrangements resulting from (uncertain scenarios of) climate change are so difficult to identify. Feminist and environmental anthropology, and the already extensive body of knowledge on rural gender relations, will provide some insights as to potential social changes in rural areas and in agriculture. Studies on the global forces at work in shaping agricultural production, rural societies, and food production are also relevant. Further work is required to analyse climate change predictions as they improve, and to consider the potential impacts of climate change on humans. Rural livelihoods and gender and power relations are embedded in social, institutional, and cultural contexts. Context-sensitive studies focusing on on-going struggles over livelihoods, status, and resources are therefore needed. These should consider the types of changes that may be in store as a result of climate change.

Poverty and environment linkages do not inevitably entail a downward spiral.

There is a great deal of variability in the ways in which local people relate to and manage their environments. Local people may respond to environmental degradation by developing technical and institutional innovations in natural resource management to reduce risks or reverse processes of degradation (Leach and Mearns 1996). Some changes are not, however, easy to detect without modern technology (e.g. the spread of disease-carrying organisms). It is important to avoid assumptions about how people will adapt to environmental change, including climate change, and the consequences of this for gender relations.

Public policies formed on the basis of the urgent need to adapt to climate change will only form a part of the actual response on the ground, since many are not enforced. Mechanisms are needed to ensure public participation in adaptive planning for climate change. Direct representation of poor people, particularly women, in developing adaptation responses is critical if such responses are to be responsive to their interests. Whilst specific actors can make decisions, act, and innovate, different visions of the future and certain courses of action are legitimised by others (Long 1997). This is true of public policy-making and has to be borne in mind in terms of the kinds of adaptations that will be legitimised in relation to climate change, and how these might in turn affect equity in gender relations. For example, public policies could rely upon coping strategies that are dependent upon women's unpaid labour if gender-awareness is lacking.

Deliberative democracy approaches, (participatory processes and mechanisms, such as citizen juries, which enable citizens to reflect upon and research issues of importance to them) have been used to encourage public debate about the consequences of complex scientific developments and political processes (e.g. genetically-modified organisms), and might be an option for increasing civil society awareness of, and engagement with, climate change.

Donors have been slow to face up to the potential significance of climate change, but this is starting to change. The World Bank, United Nations Development Programme (UNDP), Department for International Development (DFID), and the European Union (EU) have recently commissioned work on how climate change could affect the achievement of the Millennium Development Goals.[2] Since several of these goals relate to gender issues, it is hoped that the study will address gender issues. Technical research includes developing new crop varieties tolerant of salt, water, and heat stress, which could reduce women's workload (e.g. new West African rice varieties that smother weeds) (DFID 2002). Research analysing Kyoto-related projects has found that sustainable forestry, land use, and livelihood criteria need to be integrated into international carbon offset policies (DFID 2002). Gender mainstreaming should be added to that list.

58

'No-regrets' measures (providing benefits now and possibly in the future) are required. Measures are needed that promote increased resilience of poor peoples' livelihoods and that tackle gender inequality *now*, whilst increasing climate change 'preparedness' for the *future*. A great deal of work is on-going in areas such as sustainable agriculture, agro-ecology, advocacy for farmers' rights, and disaster planning, but more support for such work is required, and particularly for gender awareness to be integrated. Such measures should challenge stereotypes about gender roles, women's unpaid time, and their centrality in coping strategies, and take account of the varied and changing relationships between people, poverty, and their environments. Government and civil society capacity-building in poorer countries and vulnerable regions is urgently needed. Combined with context-specific vulnerability studies, this will assist in the identification of appropriate policy options, regional collaborations, and adaptation mechanisms. Importantly, such studies could also contribute to making visible the potential gender impacts of climate change – otherwise gender inequalities will be exacerbated.

Conclusions

The impacts of climate change on gender relations have not been widely studied to date – they therefore remain invisible. Despite the difficulties of prediction, it is clear that the impacts of climate change will be gendered, and that these require further research. Pre-existing vulnerability to natural hazards and long-term climate change means that those most at risk of, and least able to cope with, slow- or rapid-onset disasters and environmental change, are the poorest, including poor women. There are also possibilities for positive changes to occur, as we have seen in the aftermath of disasters, when women take on new roles, challenging gender stereotypes.

Public policies need to ensure that gender analysis is fully integrated to avoid exacerbating gender inequalities and to promote gender equity.

Valerie Nelson (Social Development Specialist), Kate Meadows (Gender and Development Specialist), Terry Cannon (Reader in Development Studies at the School of Humanities/NRI), John Morton (Reader in Development Anthropology and Associate Research Director), and Adrienne Martin (Social Anthropologist and Head of Livelihoods and Institutions Group) form part of the Livelihoods and Institutions Group, Natural Resources Institute, University of Greenwich, Medway Campus, Central Avenue, Chatham Maritime, Chatham, Kent ME4 4TB, UK.
Tel: +44 (0)1634 880088;
E-mail: v.j.nelson@gre.ac.uk

Notes

1 Thanks to Chris Sear, Natural Resources Institute, for his comments on draft versions of this article.
2 Correspondence with DFID adviser A. Herbert (2002).

References

Baden, S., C. Green, A.M. Goetz, and M. Guhathakurta (1994) 'Background report on gender issues in Bangladesh', *BRIDGE Report* No. 26, Brighton: Institute for Development Studies, University of Sussex

Baxter, P.T.W. (1993) 'The new East African pastoralist', in J. Markakis (ed.), *Conflict and the Decline of Pastoralism in the Horn of Africa*, Basingstoke: Macmillan/ISS

Blaikie, P., T. Cannon, I.Davis, and B. Wisner (1994) *At Risk: Natural Hazards, People's Vulnerability and Disasters*, London: Routledge

DFID (2002) 'Climate Change: How can DFID's Natural Resources Research Help?', London: Rural Livelihoods Department Research Section, DFID

Dalfelt, A. (1998) 'Climate change and sub-Saharan Africa: issues and opportunities', Newsletter 1998/10/3, World Bank, http://www.worldbank.org (last checked by author April 2002)

Delaney, P.L. and E. Shrader (2000) 'Gender and post-disaster reconstruction: the case of hurricane Mitch in Honduras and Nicaragua', draft report for LCSPG/LAC Gender Team, World Bank, http://www.worldbank.org/aftdr/ik/default.htm (last checked by author April 2002)

Dregne, H.E. (2000) 'Drought and desertification: exploring the linkages', in D. Wilhite (ed.), *Drought, A Global Assessment*, Volume II, London and New York: Routledge

Enarson, E. (2000) 'Gender and Natural Disasters', *Working Paper* 1, Recovery and Reconstruction Department, Geneva, September

Hendy, C. (2001) 'Appendix: Statistical Frequency of Drought in Northern Kenya and its Effects on Livestock Population', in J. Morton (ed.) (2001a)

IPCC (2001) 'Summary for Policy Makers. Climate Change 2001: Impacts, Adaptation, and Vulnerability', a report of Working Group II of the Inter-governmental Panel on Climate Change

Leach, M. and R. Mearns (1996) *The Lie of the Land: Challenging Received Wisdom on the African Environment*, Oxford: Heinemann/James Currey

Long, N. (1997) 'Commentary on Part I: Theoretical Reflections', in D. Goodman and M.J. Watts (eds.), *Globalising Food Agrarian Questions and Global Restructuring*, London: Routledge

MADREF/World Bank (2000) 'Programme de Développement des Zones "Bour"; Rapport Préliminaire d'Identification', Rabat: Ministère de l'Agriculture, Développement Rural, et des Eaux et Forêts and World Bank

Meadows, N. (1999) 'Second Consultancy Report: Gathering of Information on Currently On-going Activities in Pastoral Areas and Socio-Economic Study on Pastoralist Decision-Making Processes with regard to Natural Resource Management', EU Agriculture/ Livestock Research Support Programme, Kenya, Chatham: Natural Resources Institute

Mendelsohn, R. and A. Dinar (1999) 'Climate change, agriculture, and developing countries: does adaptation matter?', *World Bank Research Observer* 14(2), 2 August

Morton, J. (ed.) (2001a) *Pastoralism, Drought and Planning: Lessons from Northern Kenya and Elsewhere*, Chatham: Natural Resources Institute

Morton, J. (2001b) 'Annexe 7: Prévention et lutte contre les effets de la sécheresse', in FAO-IC, 'Programme de Développement Rural Intégré de Mise en Valeur des Zones Bour; Rapport de Préparation', Rome: FAO

PAHO (2001) 'Gender and Natural Disasters', Fact Sheet, Program on Women, Health, and Development, Washington DC: Pan-American Health Organization

World Bank (2000) 'Seeds of Life: Women and Agricultural Biodiversity in Africa', *IK Notes* 23, August, http://www.worldbank.org/aftdr/ik/default.htm (last checked by author April 2002)

Gendering responses to El Niño in rural Peru

Rosa Rivero Reyes[1]

Climatic disasters are a recurrent problem in Peru. The impacts of disasters differ between and within regions and communities. Rural upland communities, largely dependent on small-scale agriculture and natural resources for survival, are particularly vulnerable to the negative effects of extreme climate events. Government policies have not only failed to mitigate this vulnerability, but have served to exacerbate it. Women face particular vulnerabilities in the context of extreme climate events. Traditional analysis and government policy approaches have served to obscure these. This article reflects on the gender-specific lessons learned by the Centre for Andean Advancement and Development, CEPRODA MINGA, during its work with poor rural communities in the Piura region of Peru in the aftermath of the 1997-8 El Niño phenomenon. It focuses on the ways in which rural communities, and women in particular, have traditionally been excluded from policy creation, and considers how they can become influential social and political actors creating their own strategies for sustainable development and disaster mitigation and preparedness.

Disasters are a recurrent problem in Peru. Over the past five years, around one million Peruvians have been directly affected by major disasters, and perhaps the same number again have experienced the negative effects of smaller-scale events (Villarreal 2002). One of the major disasters to take place in recent years was the 1997-8 El Niño phenomenon. El Niño is a regular climatic occurrence. It takes place every five years or so, when the cold Humboldt current that flows north from Antarctica along the coast of Chile and Peru is replaced with a warmer, southern-flowing current from the tropics. This new current raises sea temperature, and causes heavy rainfall, floods, and landslides in some areas, and drought in others. The severity of El Niño's impact varies from year to year, and from place to place.

While the relationship between global processes of climate change, and specific climate events like the 1997-8 El Niño remains unclear, one of the predicted outcomes of climate change is that extreme climate events will occur with greater frequency and severity. Existing experiences of responding to climate-related disasters, particularly those amongst more vulnerable populations, can offer important lessons for informing disaster prevention and mitigation in the future.

In Peru, the impacts of the 1997-8 El Niño phenomenon were particularly severe. Over 100,000 homes were either damaged or destroyed by floods and landslides, affecting around half a million people. Three-quarters of those affected were from rural areas (Villarreal 2002).

The Centre for Andean Advancement and Development, CEPRODA MINGA, is an NGO working in the Piura region of northern Peru. This mountainous, predominantly rural area is particularly vulnerable to the negative impacts of

El Niño events. CEPRODA MINGA works with local communities to build women's and men's capacities as social and political actors, and to strengthen local institutions. CEPRODA MINGA undertakes participative planning with the region's rural communities in order to influence the formation of local, regional, and national policies for sustainable development and disaster prevention. A principal objective of this work is to develop a 'gender policy agenda' for the region, and to mainstream an understanding of gender relations into all policy formation for sustainable development.

Differential impacts of the 1997-8 El Niño phenomenon in the Piura region

The impacts of the 1997-8 El Niño phenomenon varied considerably within and between regions in Peru. While agriculture in the coastal areas of Piura benefited from improved climatic conditions, its upland areas experienced heavy rainfall causing soil, forest, and crop degradation, and leading to decreased agricultural production and capacity (Torres 1998).

Small-scale agriculture and natural resources represent the principal source of food and income for rural upland communities in the Piura region. Repeated experience of severe El Niño events over recent decades, with inadequate environmental, economic, and social recovery in-between, has diminished these communities' abilities to prepare for, and cope with, disaster. Food insecurity is an ongoing problem in the region, exacerbated, rather than created, by El Niño. The absence of an agrarian development policy with a focus on small-scale agriculture has led to food insecurity, increased rural-urban migration, and environmental degradation. In addition, before the disaster, national economic policies had favoured developments in coastal areas over those in the rural uplands. As a result, communities in upland areas found themselves in a doubly vulnerable situation at the time of El Niño. To survive in these conditions, small-scale agriculturalists have to exert considerable pressure on woods and other natural resources in order to supplement diminishing returns from agriculture. This negative cycle leaves their least powerful members increasingly vulnerable to loss, damage, and food insecurity (CLADEM 2001).

The losses and damages sustained during the El Niño event caused household income to fall dramatically during 1997, and increased the exposure of rural households to acute food insecurity. During the most critical period of El Niño, when many rural communities were flooded or cut-off, food supplies were extremely scarce, and prices increased to levels beyond the incomes of the poorest households.

Analysing the gender-differentiated impacts of El Niño

While the scientific community has developed a better understanding of, and ability to predict, El Niño events, this research has not prioritised a social analysis of the effects of El Niño, or a gender-differentiated understanding of El Niño's impacts. Where social data have been collected, these have often been aggregated in a way that obscures gender differences. The observations below have arisen from CEPRODA MINGA's work with rural communities in Piura.

Discrimination against women means that women in rural Piura typically have low access to education, specialist technical assistance, healthcare, or control over the family's productive resources. These widespread and profound inequalities put poor women (and their children) in a situation of particular vulnerability to food insecurity during El Niño. Gender inequalities in food distribution and consumption within households were common, including during periods where households as a whole appeared to have sufficient food. Widespread malnutrition also exposed women and children disproportionately to epidemics (acute respiratory and diarrhoeal infections, malaria, dengue, and cholera), which increased significantly during El Niño. Pregnant women were at particular risk from malaria, which causes serious complications during pregnancy, and other peri- and post-partum illnesses.

Increased migration of men out of the area into the coastal valleys and cities in search of employment increased the numbers of temporarily female-headed households. Female-headed households faced particular challenges in their attempts to survive the effects of El Niño. Women heads of households were typically not recognised as such by the major rural community organisations (largely led by men).[2] Equally, the increased burden of household and agricultural work placed on women in the absence of men posed an acute limitation to their ability to seek paid employment.[3]

Nonetheless, as we shall see, women in Piura were able to develop various survival strategies and capacities with which to tackle the problems generated by El Niño.

Government responses to El Niño and disaster prevention in Peru

The El Niño policy context

Peruvian regional authorities have barely acknowledged the differentiated effects of El Niño in the Piura region. This differentiation has political implications, and is connected with regional processes of development. Over the past 50 years, regional development models have favoured the development of agro-export industries (cotton and rice) in the coastal valleys, along with the development of

major oil industry and irrigation infrastructure projects. Communities in rural upland areas have largely been excluded from these processes of development, and have suffered from the absence of an agrarian development policy focused on the needs and realities of subsistence farmers. National economic policy is dependent on primary exports of minerals, fish, and agricultural products, tending to marginalise considerations of environmental protection, sustainability, or small-scale production for local consumption. The concentration of economically-important industry in the coastal region also entailed that the majority of emergency and reconstruction interventions taking place during and following El Niño were focused on coastal areas.

Understanding and responding to disasters – the failure of top-down responses

The mainstream view of El Niño events is to consider them as isolated and bounded disasters, arising from natural causes, which must be scientifically understood, predicted, controlled, and prevented using large-scale technical interventions (Wilches-Chaux 1998). The Peruvian government's response to the threat of El Niño has typically been to prioritise the construction of preventative physical infrastructure and other technical responses, and to focus this on regions of greater national economic importance. A view of disasters as isolated occurrences creates an approach to civil defence that is restricted and temporary, and prevents its institutionalisation as a part of everyday life. There has been little attempt to mobilise the population in disaster prevention. The 1997-8 El Niño event highlighted the gross weaknesses in the National Civil Defence System, as local and regional actors had practically no involvement in decision-making processes. The inevitable result of this has been the creation of a widespread sense among the population that disaster response is a matter for the state, and not for communities themselves to confront (Rivero and Cuba 2001).

A contrasting approach, taken by CEPRODA MINGA and others, is to view El Niño and other disasters as the outcome of long-term social and political processes. These disasters expose the vulnerability of people faced with environmental threats, caused by natural events or human activities. This approach accounts more effectively for the pattern of differentiated effects of El Niño, within and between communities and households. It also accounts for the long-term accumulation of vulnerabilities amongst rural upland communities that have been repeatedly affected by disasters, with little opportunity or assistance to rebuild their capabilities. The political invisibility of these communities has left them excluded from wider development processes. This, combined with repeated severe El Niño shocks, the lack of a powerful civil defence movement, and a lack of access to government emergency responses, has locked both men and women, and particularly women, into a cycle of environmental degradation of increasingly marginal lands, and resulting continual food insecurity (CEPRODA MINGA 1999).

El Niño events are an inevitable aspect of the Peruvian climate. They bring both opportunities and threats in accordance with their severity, and the geography of specific places. It is essential for communities to adapt to these in order to survive and to develop sustainably.[4] In the view of CEPRODA MINGA, this can only come about through the transformation of the social and political processes that generate disasters into processes creating sustainable development. This requires the full participation of all community members (CEPRODA MINGA 2001).

Mainstreaming gender in development work and humanitarian response

While it is clear that there have been efforts on the part of public and private institutions in Piura to incorporate a gender perspective into their work, this has commonly been treated as a technical aspect of the planning and analysis of development projects. This approach impedes the development of a better understanding of women's empowerment as a social and political process aimed at transforming the unequal relationships between men and women, within households, communities, and society at large.

Women organising locally for survival

During recent decades, women in the Piura region have been active in forming a range of women's organisations locally; these include the *Comités de Vaso de Leche* ('glass of milk committees'), *Comedores Populares* (canteens), and *Clubes de Madres* ('mothers' clubs'). All of these forms of organisation are intended to improve food security and nutrition within rural communities. Within these organisations, women members have full participatory rights to vote and voice their opinions.

During the critical period of El Niño, women leaders from Alta Piura assumed a decisive role. They took a lead in re-housing families who had lost homes, managing the distribution of emergency aid, and forming local work groups. There were many opportunities for women to demonstrate their skills as community leaders and protectors, despite their increased vulnerabilities.

However, this process of formation of women's survival organisations, and their high profile during the disaster, was only very weakly linked with wider processes of political or social empowerment of women (Rivero, Afonso, and Eggart 2002). While women were active in leading interventions locally, they were largely absent from institutions at the district, provincial, and regional levels. While they were accustomed to having voice and vote within the dedicated women's organisations, the 'principal' community organisations like the *Comunidades* and *Rondas Campesinas* (civil guards) only gave voting participation to widows, single women, and other women without a man to represent them. Although women's contributions to community survival within the emergency context were widely recognised, men held all the technical, management, and decision-making roles in the civil defence committee, the principal organisation charged with responding to the disaster.

While disasters like El Niño can offer women opportunities to assume new leadership roles and activities at a local level, the experience of the 1997-8 El Niño event showed that this increased local visibility was not translated into wider transformations in gender relations. When combined with the increased pressures and vulnerabilities that poor rural women in particular faced during the El Niño crisis, there is a real danger that women are simply burdened with extra tasks for no political or social reward. Project planners in disaster situations need to take care that encouraging women's greater participation in community-level initiatives does not load them with additional tasks and responsibilities, while failing to accord them with greater power and access to formal political bodies and national development processes.

CEPRODA MINGA realised that in the context of disasters, women's needs change. This is a result of the ways in which women's increased role as protagonists within the emergency response combines with the disproportionate risk that women will suffer chronic malnutrition or illness (see Table 1).

CEPRODA MINGA – developing strategies for civil defence

CEPRODA MINGA considers that the management of disaster risk is one of the central concerns of disaster prevention. When disasters are viewed as the result of long-term imbalances between societies and their environment, disaster risk-management also becomes a long-term commitment. This approach emphasises the management of risk and vulnerability *before* disaster occurs (Rivero and Cuba 2001). We use the term 'disaster risk-management' to refer to the technical and political capacity of women and men, and their organisations and institutions, to transform the social processes that generate disasters and convert them into processes of sustainable development. Democracy is essential to this transformation.[5]

Capacity-building with rural people to enhance their full participation in political and social processes, and particularly in the creation of regional development policy, has been the main focus of CEPRODA MINGA's work following the El Niño event. We have been able to take advantage of wider political changes in Peru that have followed the end of the Fujimori government, including a renewed emphasis on democracy and decentralisation, and the strengthening of national civil defence systems. From the outset, we knew that overcoming discrimination against women would be crucial if women, and poor women in particular, were not to be excluded from the benefits of these changes.

The CEPRODA MINGA interventions
The CEPRODA MINGA capacity-building projects were implemented during the reconstruction period following the 1997-8 El Niño, in Chalaco, a very remote mountain district in Piura. The outcomes of El Niño in this district included forest and

Table 1: *Capacities and vulnerabilities of rural women in the context of the 1997-8 El Niño disaster*

	Vulnerability	Capacity
Citizenship and social organisation	• Increased numbers of women face a double burden of responsibility, as both income-generators and carers for children and the elderly • Women frequently do not have control of resources (water, land, housing). As such, they may be limited in their ability to make decisions in these areas during an emergency. • Women have limited access to training, information, or education • Domestic violence is frequent and widespread, as a result of economic difficulties • Women often feed their families in preference to themselves and are thus at increased risk of malnutrition • Women are not able to participate in the 'principal' decision-making organisations	• Educators • Social sensitivity and capacity for solidarity • Transparent management of resources • Willingness to learn and to share
Psychological attitude	• Women perceive themselves to be dependent on their husbands • Women experience increases in stress as a result of food insecurity and epidemics • Women perceive themselves to be marginalised	• Women have a strong sense of family and community responsibilities • Women have the capacity to mobilise their organisations • Women exhibit strong pragmatism
Physical and material factors	• Pregnancy • Illness • Lactation in a context of widespread malnutrition	• Women as protectors of their community • Women search for means of survival

crop destruction, increased soil erosion, and an increase in illnesses as a result of food insecurity and shortages of clean water.

CEPRODA MINGA initiated participatory planning processes within the Chalaco communities. These processes were intended to go beyond technical discussions of disaster prevention and management to consider how communities might build new kinds of social and political relationships and institutions.

These relationships and institutions would involve all their members in decision-making and consensus-building.

Creating sustainable risk-management mechanisms among disaster-ridden communities is dependent upon 'social capital': the intricate web of social relations and networks that characterise those communities. One aim of the participatory planning processes was to identify these relationships and networks, and to enable communities to use them as a basis for the development of strong rural institutions. Another aim was to explore and value different forms of local knowledge and culture, and to consider how this 'cultural capital' might be valuable in the management of disaster risks.

The importance of mainstreaming gender relations into participatory planning

Gender perspectives were integrated into all stages of the planning process, as part of the process of valuing different kinds of knowledge and cultural and social capacities. However, we noted from the early stages of our intervention that when men and women interact, women have a tendency to subordinate their gender-specific needs and demands within the wider discussion. Through the intervention we learned that it was necessary to empower women in articulating their gender-specific needs to enable them to negotiate solutions in participatory decision-making fora.

While women have been able to participate in decision-making bodies at the level of their community or district, they have typically been absent from other, higher, levels of decision-making. We realised that local-level plans to articulate and respond to women's gender-specific needs would fail unless they were linked to higher-level changes in public policy. For this reason, we realised that, in addition to the local participatory planning processes, it would be important to create a 'gender policy agenda', with the aim of influencing and creating gender policies at the regional level. A key aspect of this is the empowerment of women to demand accountability in relation to these policies, to ensure that gender policies are not simply reduced to empty declarations of principles or tools for technical analysis, or filed away in the offices of bureaucrats.

A key part of the promotion of the regional gender policy agenda, which is ongoing, involves building on the base of women's widespread local organisational processes to create regional representatives who can negotiate gender-specific demands in decision-making fora at all levels. Currently, CEPRODA MINGA is promoting the 'gender policy agenda' at a regional level in an electoral context: for the first time, regional elections are being held in all regions of Peru. Through consultation with public and private institutions, CEPRODA MINGA is supporting a regional central-isation process for women's organisations. It sees these as as important emerging actors in the country's democratic transition.

However, we realise that we need to create a permanent regional-level gender post if the gender policy agenda is to become reality.

Increasing women's participation in local organisations

The participatory planning processes, and the development of women's capacities to articulate their strategic gender needs, are beginning to increase women's visibility in community decision-making spaces. Women are increasingly participating in decision-making spaces that would have been 100 per cent masculine just a few years ago. Men have increasingly learned to listen to and take into account women leaders' opinions as a result of the consultation and participatory processes. As a result, the women of Piura have made significant advances in participating in wider development processes, and ensuring that their needs and interests are being included in development strategies. Local systems for sustainable development and disaster prevention have been strengthened through women's participation.

From the beginning of our interventions, existing women's organisations such as the 'glass of milk' committees, mothers' clubs, and the *Rondas Campesinas Femeninas*, actively participated in meetings, guaranteeing the representation of their households and localities. In some cases, women have taken on major responsibilities, for example in the hamlet of Nogal Chalaco, where for the first time in history a woman was elected president of the Committee for Small-Scale Producers of Coffee and Sugar Cane.

There have been changes within local-level community organisations, where women have taken on more leadership positions representing women's interests, and gained voting powers and a voice in district assemblies. There has also been greater recognition of women's rights in local judicial systems. In the 'principal' local organisations like the District Assembly of the Chalaco *Rondas Campesinas*, women and men now both have a right to vote, with women also assuming some leadership roles. Where the *Rondas Campesinas* are involved in local judicial processes, there have been advances in their willingness and ability to recognise women's rights, something which has not taken place widely in the formal justice system.

In mixed organisations like the small-scale coffee and sugar-cane producers' committees, women have became more conscious of their role as producers and citizens. In Peru, as elsewhere, there is a widespread devaluation and lack of recognition of women's role in productive activities. This is considered at most as a supportive role within the traditional household division of labour. Women have traditionally considered the productive activities that they undertake to be a form of support to male producers. Typically, women who undertake agricultural activities declare themselves to be house-wives or petty traders at the time of the census. As a result, women's role in productive activities has been invisible in registers and statistics, and has not been widely recognised or appreciated.

Since the beginning of the CEPRODA MINGA interventions, women have begun to make themselves more visible within these organisations, as producers involved in all aspects of production. Increasingly, women are assuming leadership posts in the central directive, and promoting technological innovation in their small-holdings, which is in turn strengthening the capacities of new female leaders.

Rural women – from survival strategies to regional development

For rural women there has been a significant shift from focusing on survival strategies within local community groups to engaging in wider development processes. Women have moved from expressing demands linked to their practical needs, such as improvements in feeding programmes and service delivery, to making their 'strategic' gender needs and their role as social actors increasingly visible in local consultation processes. Today, women are demanding to be considered as workers equal to men, within development programmes; they express the need to build capacity in the exercise of their rights in order to confront situations of gender violence and abuse, and to challenge instances of unjust rulings and decision-making by authorities in favour of men. According to statistical data from the area, boys and girls have now got equal access to public education, which is an important change compared with 1990 when girls had very limited access.

Conclusions

Sustainable development in Piura will be a long-term process. Similarly, we realise that there is a long way to go in achieving real equality between women and men in all areas. Nonetheless, as a result of the CEPRODA MINGA interventions, a better understanding of disaster prevention has developed within the Chalaco communities. Where, previously, the idea of rural development was associated with physical infrastructure and centralised decision-making, CEPRODA MINGA's interventions have contributed to the creation of a widespread new understanding of how people can develop their own capacities to transform their situation. At an assembly meeting, both male and female participants spoke of how they now felt able to talk directly with authorities, whereas before they went through intermediaries. This, they felt, had enabled them to claim their rights as citizens and take their proposals to larger political fora that have hitherto been considered as excluding and ignoring rural people's concerns. Importantly, the participative planning process has enabled rural people – women and men – to create and promote their own proposals for democratic government and local development in a context of decentralisation.

Rosa Rivero Reyes is the Executive Director of CEPRODA MINGA. Contact: Centro de Promoción y Desarrollo Andino CEPRODA MINGA, Residencial Grau F-201, Piura, Peru. Tel: +51 (0)74 309701; Fax: +51 (0)74 309703; E-mail: ceproda@qnet.com.pe

Notes

1. This article has been translated from Spanish and adapted by Kate Kilpatrick.
2. These organisations typically only recognise women who do not have a man to represent them, such as single women and widows. When a woman's husband is absent from the community for migratory work or some other reason, she is not usually recognised by these organisations as the acting household head.
3. Rural women who migrate to the larger towns in the region usually obtain income through domestic work or through petty commerce. During El Niño, these opportunities decreased considerably as households across the region were affected by the disaster.
4. In fact, communities in Peru were previously better-adapted to El Niño events than they are today. Agricultural changes and humanitarian aid packages have created widespread dependence on external inputs, displacing traditional native plant varieties (for example, the yacon [*Smallanthus sonchifolius*], and native bean and potato species) which are often better adapted to the climatic conditions.
5. The communities that displayed a better organisational response to El Niño were those where participative processes and grassroots NGOs were in existence long before the disaster. These institutions and communities appeared to emerge strengthened rather than undermined from the El Niño emergency and reconstruction period. Those communities and institutions with scant experience of popular participation typically fell prey to welfarism and clientelism.

References

CEPRODA MINGA (1999) 'Plan Estratégico para el Desarrollo Sostenible y la Gestión de Riesgos de Desastres de la Microcuenca Nogal Chalaco', Piura, Peru: CEPRODA MINGA

CEPRODA MINGA (2001) 'Plan Estratégico del Distrito de Chalaco para el Desarrollo Sostenible y la Gestión de Riesgos de Desastres', Piura, Peru: CEPRODA MINGA

CLADEM (2001) 'Perú Diagnóstico de la Situación de los Derechos Sexuales y los Derechos Reproductivos 1995-2000', Lima: CLADEM Peru

Rivero, Rosa and Severo Cuba (2001) '"Primero es la Gente": Prevención de Desastres, Borrador de Documento de Sistematización', Piura, Peru: CEPRODA MINGA

Rivero, Rosa, Klara Afonso, and Luisa Eggart (2002) *Guía Metodológica: Construyendo una Agenda Politica de Genero en la Region Piura: Los Derechos Humanos de las Mujeres*, Piura, Peru: CEPRODA MINGA

Torres, F. (1998) 'Efectos de "El Niño" en Cultivos y la Productividad Primaria Vegetal en la Sierra Central de Piura', paper presented at a workshop on '"El Niño" in Latin America: its Biological and Social Impacts. The Basis for Regional Monitoring', 9-13 November 1998

Villarreal, L.A. (2002) 'Vulnerabilidad, desastres y desarrollo en el Perú', in *Pobreza y Desarrollo en el Perú*, Lima: Oxfam GB

Wilches-Chaux, Gustavo (1998) *Guía de la Red para la Gestión Local del Riesgo*, Lima: ITDG Peru and Red de Estudios Sociales en Prevención de Desastres en América Latina

The Noel Kempff project in Bolivia: gender, power, and decision-making in climate mitigation

Emily Boyd

A focus on land-use and forests as a means to reduce carbon dioxide levels in the global atmosphere has been at the heart of the international climate change debate since the United Nations Kyoto Protocol was agreed in 1997. This environmental management practice is a process technically referred to as mitigation. These largely technical projects have aimed to provide sustainable development benefits to forest-dependent people, as well as to reduce greenhouse gas emissions. However, these projects have had limited success in achieving these local development objectives. This article argues that this is due in part to the patriarchal underpinnings of the sustainable development and climate-change policy agendas. The author explores this theory by considering how a climate mitigation project in Bolivia has resulted in different outcomes for women and men, and makes links between the global decision-making process and local effects.

By and large, climate mitigation projects have been informed by Western ideas of science and development, and predominantly driven by the 'masculine' interests of forestry, accounting, agriculture, and policy-making. This article aims to articulate some of the concerns arising from this agenda, and demonstrates how a patriarchal system of decision-making exists at all levels, from global decision-making frameworks to the local implementation of climate mitigation projects. The predominant decision-makers at all levels of decision-making are men: bureaucrats negotiating on behalf of their governments; NGO representatives; extension workers; and decision-makers in local organisations. Concern about the lack of a gender discourse within debates about climate change debate has been raised by writers such as Vandana Shiva (1988). When we discuss global carbon-trading by

planting or conserving trees, we are engaging in a debate driven by men, who are biased towards providing technical solutions to the climate-change problem, and who have little understanding of, or regard for, the concerns or interests of women.

This article is based on my doctoral research, which was undertaken on one of the world's largest UN carbon sequestration pilot projects: the Noel Kempff Climate Action Project in the Bolivian Amazon. The research took place between March and September 2001. I used a qualitative approach, drawing on informal interviews, participant observation, rapid rural appraisal, participation in meetings, and an evaluation workshop. The key research objective was to establish how carbon sequestration projects contribute to local sustainable development, and to assess the compatibility of local institutional arrangements with the highly politicised goals and rules of external interventions. The research focused predominantly on the social, institutional, and development contexts within which the project was established, i.e. the framework within which the project took place.

In the process of the research, it became clear that the project was weaker in some areas than others. These weaknesses can be understood within a framework of practical and strategic gender needs (Moser 1993), and have been highlighted as important across a range of projects and programmes (Regmi and Fawcett 1999; Sardenberg *et al.* 1999). 'Practical gender needs' refer to the immediate necessities that women perceive themselves as lacking in a specific context, which would enable them to perform the activities expected of them: for example, a health post, vegetable gardens, or a water pump. 'Strategic gender needs', in contrast, refer to that which is necessary for women to change their status in society. These might include: access to and ownership of land or other property, control over one's body, equal wages, or freedom from domestic violence.

The Noel Kempff project has been operating for five years. During this time, opportunities have been provided for the participation of both men and women. It would be fair to say that women have successfully participated in some aspects of the project, fulfilling practical gender needs such as trying new varieties of legume crops or accessing credit, but that they have predominantly been recipients of charity. Addressing strategic needs has not been a key factor in the project design and implementation. It was evident during the research process that a gender perspective was lacking. In particular, a perspective was needed which recognises the difference between practical and strategic gender needs, and the existence of gendered institutions, power structures, and hierarchies. I consider these concepts in more detail in the following section.

Embedded patriarchy: the absence of a feminist perspective in existing climate change frameworks

Mitigation through forest management

Growing international concern about climate change has resulted in a number of United Nations agreements, including the United Nations Convention on Climate Change, and the Kyoto Protocol, which came out of it. The Protocol was created in 1997, with the aim that it should be ratified by 2002. It is the first legally binding global commitment to tackling developed countries' greenhouse gas emissions. A key greenhouse gas is carbon dioxide (CO_2), which is chiefly emitted into the global atmosphere through the burning of fossil fuels (industrial emissions are estimated to account for five billion tonnes of carbon emissions, and deforestation 1.6 billion tonnes).

Tropical forests absorb carbon dioxide through the process of photosynthesis. This knowledge led to the inclusion of forest 'sinks'[1] within the policy debate, and initiated a project pilot phase that would test the possibility of mitigating (offsetting) CO_2 emissions through sustainable forest management, conservation of forests (avoided deforestation), or planting trees (afforestation and reforestation). These projects aim to reduce pressure on forests by improving the management of land and forests. The Protocol formally establishes the possibility of reducing carbon emissions through the *flexibility mechanisms*, which allow trading of carbon dioxide on an international market through developed country investments (primarily industry) in carbon mitigation projects in developing countries, such as clean energy substitution in China or forest projects in Brazil. There are also informal discussions ongoing about an alternative carbon market, which would trade carbon from a wide variety of forestry projects. Forest-based CO_2 mitigation projects were established to provide 'win-win' solutions under the umbrella of sustainable development.

'Masculine' bias of mitigation approaches

The new and emerging field of climate change mitigation through project activities in developing countries is based on modern scientific concepts. According to Vandana Shiva, the approach is based on a world-view that supports, and is supported by, the socio-economic and political systems of Western capitalist patriarchy, which dominate and exploit nature, women, and the poor (Shiva 1988). This section gives a feminist perspective on the climate change debate at global and local levels, and uses a 'feminine', human-centred, and rights-based approach. If we look at the rules, norms, and aspirations of the institutions which are involved in the debate through a *feminine* framework, we can understand the inequalities which exist in global decision-making and power structures, and see how the legacies of colonialism still shape the main institutions in developing countries today.

Historically, the Western mode of development has reinforced a patriarchal style of decision-making. This is reflected in the predominantly technical approach of employees of conservation and sustainable development projects. Denton (2001, 1) notes that, 'The climate change debate is an indicator of how gender issues tend to be omitted, leaving room for complex market-driven notions equated in terms of emissions reductions, fungibility, and flexible mechanisms.' These highly technical terms reflect the extent to which complex issues are glossed over and simplified within global responses to climate change. Solutions to the problem might instead be found through focusing resources on understanding how climate change will affect women and men differently, and what measures are necessary to ensure adaptation.

The UN, a key decision-making body in climate change issues, has a male-dominated, hierarchical structure. Denton observes how an overall assessment of the climate change debate to date shows that women are absent from decision-making processes, and that decision-making and policy formulation at environmental levels, within conservation, protection, rehabilitation, and environmental management, follow pre-dominantly male agendas (Denton 2001, 1). Not only are the bureaucrats representing their nations predominantly men, but more importantly, the underpinning approach is 'masculine'. The power of an alternative 'feminine' approach to environmental governance and management should not be underestimated. Increased gender aware-ness within global decision-making on climate change should allow for inclusion of different socio-economic groups, rather than professionals only, and should encourage the participation and repre-sentation of those women most vulnerable to climate change (*ibid.*). In her foreword to *Staying Alive* (Shiva 1988), Rajna Kothari suggests how such an approach might look. 'The struggle for femininity is a struggle for a certain basic principle of perceiving life, a philosophy of being. It is a principle and a philosophy that can serve not just women but all human beings. Femininity by definition cannot and should not be a limiting value but an expanding one – holistic, eclectic, trans-specific and encompassing of diverse stirrings.' (Kothari 1988, xiii)

A key characteristic of international-level interventions is the exclusion of social groups, such as indigenous groups and women, from decision-making. At inter-national climate negotiations, we still see disparities between rich and poor nations, women and men, NGOs and government policy-makers. The balance of women and men directly taking part in the decision-making remains an issue to contend with, as exemplified at a recent UN climate meeting, when the numbers of male to female professionals elected to the executive board to oversee future forestry and energy projects had a ratio of 11:1.[2] This balance is replicated down the chain, from policy to project.

Within carbon mitigation projects, the inclusion of a feminist analysis would assist in the pursuit of the 'win-win' situation that scientists, policy-makers, and NGOs consider forest mitigation projects to have the potential to exemplify. In addition, matters of participation, access to information, and control over decision-making are also important.

Case study: the Noel Kempff Climate Action Project

The Noel Kempff Climate Action Project was established in 1996 in Bolivia, in the region of San Ignacio de Velasco, Santa Cruz. The project's primary objectives were to purchase logging concessions from companies and thereby expand the Noel Kempff National Park to 1.5 million hectares (almost double its original size) to meet conservation aims and earn carbon credits. The project also aimed to contribute to local development benefits through improved local agricultural and forest management practices, to stimulate employment, and to obtain 400,000 hectares of communal land for three key communities (Florida, Porvenir, and Piso Firme). These are predominantly Chiquitano indigenous communities,[3] of approximately 2000 inhabitants. Funding was obtained primarily from an electrical utility company in the USA, with some financial support from the Nature Conservancy (an international NGO), and Fundación Amigos de la Naturaleza (a local NGO). The government of Bolivia was also closely involved as a broker and partner with the private sector for the carbon credits accrued from the project.

The project has a four-component structure, comprising forestry, agriculture, conservation, and community development. Forestry activities entail monitoring CO_2 fluctuations inside and outside the park, establishing a forest management plan, and ensuring community participation in its implementation. Agricultural activities include land-use planning, testing new crop varieties, and establishing model agro-forestry farms. Conservation activities predominantly involve eco-tourism in one community, and some minor sales of handicrafts to tourists. These activities overlap with the community development programme, *Apoyo Communitario* (APOCOM).

To summarise, the key characteristics of the project are:

- a primary focus on land-use, land-use change, and forestry;
- the aim of reducing CO_2 emissions with additional biodiversity benefits;
- sustainable development objectives, manifested through locally-focused project activities.

Assessing the project's impact on gender relations

Gender-based inequalities in employment opportunities

The forestry programme provided short-term employment for between 30 and 50 men from the local communities, to establish forest inventories and plant nurseries. They received a salary of up to US$6 per day for their efforts. A small number of women were employed to cook for the forestry workers. The men received a salary for acting as community technicians and providing technical training, and were part of the technical team that managed and drove the community forest concessions in the land titling process (which is still ongoing). There was some effort to encourage the presence of women in the forestry team, but this did not succeed. Women are rarely present at forestry programme meetings and workshops (personal observation).

As part of the conservation activities, approximately 30 men were employed annually for a number of weeks to clear the roads into the national park, and two women from one village were employed as cooks when tourists came to the main park camp. The park also employed six or seven local men as park guards. These men earned up to US$100 per month, which was a very reasonable local salary.

Aside from the female cooks, one female forestry consultant involved in training the forestry team, and the female co-ordinator of the eco-tourism activities, all the technical staff directly involved in the project were male. Although the work provided much-needed income and status to these community members, the majority of women were not able to benefit directly from this.

Gender inequalities in power and decision-making

The NGO and National Park directors were both men, and were overseen by government technicians based in La Paz, who were all men. At the local level, the project has successfully reinforced the power of traditional councils (*cabildo*) by entrusting them with the land titling process. These are all-male councils consisting of a headman (*cacique*) and a council of 11 men, each with a different task. An attempt was made in one community to encourage the participation of two women in the *cabildo*, but they didn't stay for long, saying, 'We got fed up with the meetings, and you know what men are like – we didn't enjoy it.' The project has also assisted the creation of the region's own indigenous organisation CIBAPA (*Central Indigena de Bajo Paragua*), with a male president and his male assistant.

At the community level, power lay with community members who had influence, financial resources, or ties to the old patron and the headman. The other key groups in the community were the fishing and cattle committees. There was a woman mayor in the village, whom project staff said had done a lot for women. Her political appointment by the municipality, however, had created a divide between political factions in the community, so there was little identification with her as a role model (personal observation).

There was a clear distinction between project activities associated with women and those associated with men. Women primarily associated themselves with activities linked to meeting practical gender needs, which had been designed by planners with women in mind. Women generally spent their time in the fields harvesting maize or rice, collecting firewood and medicinal plants, and growing fruit trees and vegetables in their homesteads. They did not work in the sawmills, extract timber, or hunt, and rarely fished. All these are activities typically associated with men.

Neither in the context of the project, nor at community meetings, was there any evidence of a focus on the socio-political and economic roles of women in decision-making, or their relationship with their environment. For example, there are no women on the national park management committee, which is an important forum for participation in decisions on the future of the park, land title, and other related activities. Public meetings were often dominated by a small number of men. During a discussion on the communal water pump in one village, a problem that women had voiced informally about livestock dirtying the water source was not voiced in the general discussion. Instead, there was a male-led discussion about credit for cattle and a boat, and the issue raised by the women was pushed down the list of priorities.

Observations such as this raise concerns about the prioritisation of issues, the action subsequently taken, and the invisibility of women and their interests within these prioritisation processes. Women would often comment, 'My husband knows about these issues – he attends meetings.' Lack of voice was something that more marginalised men in the communities also experienced. In the smaller communities, women spoke more openly in public, yet still required a great deal of encouragement to speak openly about what they wanted from the project and in their lives more broadly.

Meeting women's needs but not advancing their interests

The project, like many development projects, focused on women's practical gender needs, such as health, education, income-generation, and food production, and neglected the strategic gender needs that could empower women, challenge the existing gender division of labour, and bring about greater gender equality (Momsen 1991).

One of the most appreciated project activities, according to many women, was the presence of a doctor, and the access to flying doctors in an emergency. However, they recognised that the emergency flights were unsustainable, and that the doctor was a temporary presence, which would last only as long as the funds were provided by the charity. Across the communities, I observed that project activities focused on infrastructure provision, such as a health post and an improved school building. Medicines were scarce: women generally obtained credit to purchase medicines, and had difficulty repaying.

At the time of the research, national economic crises were hitting these remote communities hard, and their usual employment opportunities with the sawmills in the region were scarce. The wet season proved a difficult time for communities in finding income and resources, such as fish, and the poor conditions of the roads isolated the communities for several months of the year. Alcohol consumption and drunkenness among men were a common sight within the communities (personal observation).

The main communal activity that women tried to engage in was working on the vegetable plots in each community. The NGO brought seeds for the women to use communally in their 'mothers' clubs' (*clubes de madres*), but they were not successful. Women commented that, 'The communal gardens have failed because we don't like working in groups, we fell out over who was taking gains from the garden and now people plant seeds in their individual gardens.' Or that, 'People stole from the communal garden, that is why it failed.'

The agroforestry farming initiative was only taken up by a small number of families. Many young men remained convinced that large-scale cattle ranching was the solution to their poverty. There were no female-headed households involved in the establishment of a model farm, although they would have benefited from the credit of a bull or cow for the production of milk for their children. Although this initiative was open to all community members, the model plots were very labour-intensive and would have required female-headed households to pay labourers or rely on charity from other male family members to clear plots of land.

Some women took up small income-generating activities, such as manufacturing chicken coops, and bread making. Those involved expressed pride and satisfaction at having their own activities, but the management of funds and repayment rates was less successful. In one community a large number of women were involved in a palm-canning factory, which received indirect contributions from the project.

A lost opportunity for women's participation and empowerment

If we are to ensure sustainable development associated with the interventions of these multi-component projects, they must address the strategic interests of women. Townsend et al. (1999) suggest that these interests include political strengthening, ensuring gender equity in access to education, ending gender violence, decreasing maternal mortality and morbidity, and ending the economic inequalities between women and men that leave women bearing the brunt of poverty.

The project's enforcement of existing social structures and wide reliance on traditional norms of decision-making has weakened women's ability to participate within or influence it. The project could have benefited from strengthening women's groups and addressing women's strategic interests. For example, discussion groups around the issues of women's social and political participation might have created a new dynamic. This idea is inspired by a one-day trial workshop on sexual education for young people, which attracted almost all of the adults in two communities. This demonstrates the widespread interest that exists in a subject that is commonly considered to be 'taboo' in the communities (personal observation).

In the final participatory evaluation of the project, few women were involved from each community, despite efforts to timetable the sessions around women's household duties. Education was high on women's list of priorities, and the project provided small grants (US$100) per family for school tuition for boys and girls. However, concerns were raised about 'girls going away to the city and coming back with a swelling belly'.

Improving women's access to information

Access to information, and what people do with that information, is an important aspect of empowering marginalised sectors within communities. In one community, many women living in the poorer areas noted that they did not attend the technical or other meetings, but that their sons or husbands might have. However, information is not consistently disseminated within the household or at the community level. A number of women pointed to the fact that their husband knew about the project but that they were not told about what went on at meetings. In this community, men were also called upon to help answer questions about the types of project activities going on in the community. In the smallest of the three communities, women noted that they attended all types of meetings, and 'learned a lot from them'. Where the women participated more in the meetings and took a greater interest, they were also the most outspoken in public meetings. The results suggest that although there are differences in the levels of information flow between different communities, across the board it is fair to say that the flow of information between the project, the park, the higher-level decision-makers, the CIBAPA, and the women in the communities could be significantly improved.

Conclusions

In conclusion, if there are to be 'win-win' solutions linking the poverty, deforestation, and climate crises in climate mitigation projects, these projects will have to take into account cross-cutting issues, such as the fact that the current framework is predominantly managed and implemented by men at all levels; that male-dominated social organisation is reinforced by Western scientific and development approaches; and that project activities are predominantly targeted towards men, or towards women's practical rather than strategic gender needs.

We can begin to 'bundle' these issues together. One bundle is concerned with equity (including the realisation of women's strategic needs), and another with the strengthening of women's role in local governance, participation, and institutions. The technical and top-down nature of projects can be problematic, preventing the development of participatory and inclusive structures. Therefore, design, implementation, and monitoring should be considered from a gender perspective to include men's and women's needs and interests. Challenges ahead include the questions of how local institutions and organisations can ensure that women are incorporated into the decision-making framework and consider strategic gender needs, and, at the higher decision-making level, of how to incorporate a gender analysis within climate change frameworks.

At the community level, we need to consider how to raise the issues of strategic needs through gender workshops. In the example cited in this paper, the local organisation CIBAPA has an important role in ensuring that it encourages women onto its committee. At the technical level, female foresters and extension workers should be encouraged to work with the communities. At the international policy-making level, gender perspectives should be incorporated into the culture of decision-making, which should include a broad and inclusive perspective on the one hand, and mechanisms for the inclusion of women in decision-making on the other.

Emily Boyd is a doctoral student at the School of Development Studies at the University of East Anglia, Norwich. She is researching social and economic aspects of carbon sequestration projects in Bolivia and Brazil. She also works as a writer/editor for the Earth Negotiations Bulletin. E-mail: e.boyd@uea.ac.uk

Notes

1. Terrestrial 'sinks' refer to carbon absorption outside of intentional human action. In particular, carbon uptake from non-managed terrestrial areas is thought to be the result of three processes – increased CO_2 fertilisation, increased nitrogen deposition, and impacts attributed to a changing climate. http://www.wri.org/climate/sinks.html
2. Personal observation from the Seventh Conference of Parties (COP7) meeting in Marrakech, November 2001.
3. For more information about Chiquitano culture and livelihood strategies, see G. Birk (2000) *Owners of the Forest: Natural Resource Management by the Bolivian Chiquitano Indigenous People*, Santa Cruz, Bolivia: APCOB/CICOL

References

Denton F. (2001) 'Climate change, gender and poverty – academic babble or realpolitik?', *Point de Vue*, 14, Dakar: ENDA

Kothari, R. (1988) 'Foreword', in V. Shiva (1988)

Momsen, J.H. (1991) *Women and Development in the Third World*, London: Routledge

Moser, C. (1993) *Gender Planning and Development*, London: Routledge

Regmi, S.C. and B. Fawcett (1999) 'Integrating gender needs into drinking-water projects in Nepal', *Gender and Development* 7(3): 62-72

Shiva, V. (1988) *Staying Alive: Women Ecology and Development*, London: Zed Books

Sardenberg, C., A.A. Costa, and E. Passos (1999) 'Rural development in Brazil: Are we practising feminism or gender?', *Gender and Development* 7(3): 28-38

Townsend, J. *et al.* (1999) *Women and Power: Fighting Patriarchies and Poverty*, London and Oxford: Zed Books and Oxfam

Reducing risk and vulnerability to climate change in India:
the capabilities approach

Marlene Roy and Henry David Venema

This paper argues that the ability of women to adapt to climate change pressures will be enhanced by using the 'capabilities approach' to direct development efforts. By using this approach, women will improve their well-being, and act more readily as agents of change within their communities. This argument is supported by previous research on gender and livelihoods, and a study conducted in rural India. Examples are based on the experiences of poor, rural women in India, who are particularly vulnerable to climate change impacts. Their survival is dependent on their being able to obtain many essential resources from their immediate environment. Yet these women lack many of the requirements for well-being, such as access to healthcare, literacy, and control over their own lives. Gaining these would reduce their vulnerability to their changing environmental circumstances.

The need to respond and adapt to climate change has become widely recognised, and people will have to deal with its impacts, with or without the help of government. The roles and activities of women and men are socially constructed, and gender-differentiated. Climate adaptation and mitigation strategies need to appreciate the different realities of women and men, in order to identify positive solutions for both.

As Amartya Sen and others have shown, poor rural women in India generally have fewer rights and assets than men. They experience inequalities in such areas as healthcare and nutrition; are more likely to suffer sex-selective abortion or infanticide; are less likely to receive an education; have lower access to employment and promotion in occupations; lack ownership of homes, land, and property; and take disproportionate responsibility for housework and child-care (Patel 2002). This asymmetrical division of labour, rights, and assets leaves women more vulnerable to – and less able to cope with – the additional stress and deprivation brought about by climate change.

The situation of poor rural women in India

Rural men and women in India are historically bound to its agrarian landscape, with which they have co-evolved throughout centuries of change. Today, unprecedented challenges, including a growing population, environmental hazards in the form of climate change and land degradation, and the globalisation of markets, are driving the need for fundamentally different social arrangements.

The millions of rural income-poor of India, of whom 50 per cent are concentrated in the states of Bihar, Madhya Pradesh, and

Uttar Pradesh, are caught in the middle of this sea-change without compass or rudder (UNDP 1997, 51). Since 1950, they have been rocked by the privatisation of communal land, the Green Revolution, and the introduction of often expensive agricultural technologies, and have been pushed onto marginal land, resulting in decreased yields and increasing out-migration to non-farm employment, particularly amongst men.

Marginalisation of poor rural women

Increasingly, women are sustaining their livelihoods as farm labourers rather than as cultivators, with their knowledge and labour largely marginalised as a result of mechanisation and other technical interventions, which they are traditionally excluded from using. In addition, their workload has increased, as the switch to high-yielding varieties of grains has created fewer crops and animal wastes for animal fodder and household fuel, the provision of which is largely the domain of poor peasant and tribal women (Venkateswaran 1995 Agarwal 1997). Moreover, the traditional usufruct rights that women held to community land were lost after land reforms, thus denying them access to these lands where, 'the landless and landpoor [had] procured over 90 per cent of their firewood and satisfied 69-89 per cent of their grazing needs' in the 1980s (FAO 1997; Agarwal 2001, 1625).

Women's labour linked to household welfare and income

Of the total Indian female work force, 89.5 per cent works in rural India, and contributes extensively to household welfare and income (FAO 1997). According to Venkateswaran, women are estimated to contribute on average between 55 to 60 per cent of the total labour of farm production (Venkateswaran 1995, 20). They often start contributing to household economic activities before they are 15, with some putting in a full day's work by the time they are ten. They undertake the bulk of work necessary to maintain the home, contribute manual labour to the cultivation of plots, and care for farm animals (Venkateswaran 1995, 24). There is some variation within India, however. For example, in both hill and mountain regions, and in arid and semi-arid areas where forests have disappeared and agriculture remains poor, women spend between six and ten hours daily collecting the resources they need to meet their basic survival needs (Centre for Science and Environment 1999). Those in the rich plains areas, where forest biomass has been replaced by agriculture biomass, spend less time on these tasks, though poor women in these areas who don't own land or whose landholdings are slight, find themselves at the mercy of major landowners to meet their fuel and fodder needs (*op. cit.*).

High female illiteracy rates

Rural women have few options, especially with the loss of usufruct rights to community land. Education, which could increase their choices and opportunities, remains limited or non-existent. Even though there has been an increase in the literacy rate for Indian women overall, over 161 million rural women (approximately 70 per cent) are still illiterate (Government of India 2001). While many children attend school until the age of ten, girls usually drop out earlier to help at home. In addition, rural Indian women have little power within the household, and their contribution, especially in family enterprises, is often hidden from public awareness (Simmons 1997). This lack of power extends beyond the family, as women rarely participate in community-level decision-making, and are consequently less able to act as agents of change to better their situation.

The impacts of climate change on the rural poor

Low-caste, tribal, and poor rural women, dependent as they are on their natural environment for water, fuel, fodder, and food, are immediately and adversely affected by all forms of environmental degradation, including climate change impacts. The Intergovernmental Panel on Climate Change (IPCC) considers India, with its large, agrarian population, to be acutely vulnerable to the impacts of climate change, and recent extreme weather events such as the cyclones in Orissa in 1999, and the severe drought in northern and central India in 2000, support this view. In addition, a 1998 World Bank report on the impacts of climate change on Indian agriculture maintained that these impacts would be region-specific, and could be significant for poor people living on marginal land.

While severe weather events such as cyclones, monsoons, and drought cannot be directly attributed to climate change, they do, nevertheless, illustrate the very real and probable impacts of climate change on the rural poor. The drought in Orissa, for example, forced many small and marginal farmers to give part of their landholdings to moneylenders, with unofficial estimates indicating that another half million people were forced into distress migration.

According to a briefing paper presented to the Indian Parliament by the Centre for Science and Environment in 2000, climate change manifestations in India will include increased temperatures, sea level rise along coastal regions, changes in monsoon rain patterns such as a decline in summer rainfall, increased flooding in the Himalayan catchment, and water resource problems in arid and semi-arid regions (Agarwal 2000; IPCC 2001). These impacts will affect agriculture and forestry, as well as human health. Agriculture, in particular, will ·perience decreased yields, as crop cycles ·ten (for rice between 15-42 per cent

and for wheat between 25-55 per cent), rainfall decreases, and conditions more conducive to pest infections, are created by rising temperatures. Consequently, researchers have made conservative estimates that farm incomes will decrease by 8.4-12.3 per cent (Sanghi 1997; Kumar and Parikh 1998).

A decline in farm-level income alone will have deleterious effects on the rural poor, particularly women, who are among the lowest-paid agricultural labourers (Venkateswaran 1995). In addition, women whose livelihoods depend on cultivating small plots and gathering fodder and fuel will be even more vulnerable as climate change advances, as they do not presently have access to the necessary resources or social status within households and communities (Adger and Kelly 1999). What can be done to reduce their vulnerability and help them adapt to their changing circumstances?

Increasing capabilities and reducing risk

According to Amartya Sen, there are five instrumental freedoms that, if present, and if women have access to them, will provide opportunities for women to act in their own self-interest and reduce their vulnerability. Access to these instrumental freedoms, namely political freedom, economic facilities, social opportunities, transparency guarantees, and protective security – is necessary for women to gain a better quality of life and acquire the capabilities they need to act as their own agents of change (Sen 1999).

Commonly referred to as the 'capabilities approach', Nussbaum (2000) describes this approach as, 'an approach to the priorities of development that focuses not on preference-satisfaction but on what people are actually able to do and to be'. Central to this approach is the idea that freedom is more than citizens having rights 'on paper': it also requires that citizens have the resources to exercise those rights. Thus, the capabilities approach goes beyond asking about satisfaction of people's preferences to ask what women's opportunities and liberties actually are, as well as how the available resources work or do not work in enabling women to function.

According to Sen, this 'capabilities approach' to development has considerable potential for enabling and empowering poor rural women.

> 'These different aspects (women's earning power, economic role outside the family, literacy and education, property rights and so on) may at first sight appear to be rather diverse and disparate. But what they have in common is their positive contribution in adding force to women's voice and agency – through independence and empowerment.'
> (Sen 1999, 191-2)

Development in Kerala and the capabilities approach

The state of Kerala in southern India provides some insight into the usefulness of Sen's approach, as several aspects of Kerala's development path bear similarities to those advocated as part of the capabilities approach. Despite its low per capita income, Kerala is notable because it has the lowest birth rates, highest literacy rates, and longest life expectancy in India and, hence, is a low-consumption economy that delivers a high quality of life. Civil rights campaigns and caste reforms began in Kerala during the nineteenth century. Equitable access to education spread in the early twentieth century, and again in the 1960s, and the success of a campaign for universal literacy resulted in the newly literate writing letters to government offices demanding better services such as paved roads and hospitals (McKibben 1996).

Affordable healthcare is also widely available in Kerala, along with nutrition programmes. In addition, there appears to be much less gender discrimination, and a robust media and political structure.

Recently, the government started a 'land literacy' programme known as the 'People's Resource Mapping Program', in which local villages map their local resources. These community maps are then combined with scientific maps to guide local environmental and social planning, with villagers taking and implementing the decisions (McKibben 1996).

Village women as agents of change

The provision of Sen's five instrumental freedoms is, however, dependent on cultural norms and rules that are manifested in a myriad of ways, including through roles and responsibilities within families, and through policies, practices, and legislation at community and state level. In India there exist numerous formal and informal arrangements and institutions that shape the different capabilities of men and women. One well-known example is the caste system, which is still prominent in many areas of rural India, and which greatly influences individuals' access to rights. For example, people from scheduled castes form 'the weakest economic segment of rural society with limited access to education and financial institutions, and little effective voice' (Simmons and Supri 1997, 311).

Informal institutional reform, whereby individuals at the community level become agents of change, appears to be a good option for increasing the ability of the poor to adapt to climate change. Chopra and Duraiappah (2001) indicate how vested interests work to prevent institutional change. They argue that the best development approach is through improved environmental and land management in communities, based on Sen's concept of five freedoms, which challenges the status quo. Two case studies conducted in Bihar and Rajasthan indicate that this type of development can be successful (Chopra and Duraiappah 2001). In Bihar, an informal institution called 'Chakroya Vikas Pranali' was formed to negotiate a set of rules to govern the use of local land and water resources. The success of Chakroya Vikas Pranali was attributed to transparency in decision-making and sharing benefits, risk minimisation, and increased protective security through the distribution of economic and social opportunities amongst individuals over time.

Certainly, better environmental management at the community level is seen by many experts as essential to efforts aimed at minimising climate change impacts. However, women may not benefit from such community-led change unless these local and informal institutional arrangements are shaped by the specific and often different needs, roles, and responsibilities of men and women. Research conducted by Agarwal (2001) on participation in joint forestry management projects indicates that while women may be active in all-women community groups, their participation in other community-based organisations is generally low. In some cases, women were actively excluded by men even though spaces were reserved for women on the local councils. This lack of participation by women indicates that political freedom, one of Sen's five freedoms, is not generally available to women, thus restricting what they are able to do and to become. Moreover, the absence of political freedom is critical, as it is a prerequisite for many of the changes necessary for women to take an active part in shaping rural development that meets their needs (Chopra and Duraiappah 2001)

Conclusions

By using the capabilities approach to direct land and environmental management changes in communities, the well-being of the rural poor can be improved. This has the potential to go a long way towards reducing their vulnerability to the risks of climate change. Poor rural women, who are already among the most vulnerable, must be specially considered in such development efforts, however, and their right to participate in decision-making must be promoted and protected.

Marlene Roy is a researcher on gender and sustainable development at the International Institute for Sustainable Development, 161 Portage Ave. E. – 6th Floor, Winnipeg, Manitoba R3B 0Y4, Canada.
Tel: 1 (204) 958 7724; E-mail: mroy@iisd.ca

Henry David Venema is a research officer at the International Institute for Sustainable Development, 161 Portage Ave. E. – 6th Floor, Winnipeg, Manitoba R3B 0Y4, Canada.
Tel: 1 (204) 958 7706; E-mail: hvenema@iisd.ca

References

Adger, W.N. and P. Kelly (1999) 'Social vulnerability to climate change and the architecture of entitlements', *Mitigation and Adaptation Strategies for Global Change* 4(3-4), 253-66

Agarwal, A. (2000) 'Climate Change: a Challenge to India's Economy. A Briefing Paper for Members of Parliament', occasional paper, Centre for Science and Environment, New Delhi: CDE, http://www.cseindia.org/html/cmp/cse_briefing.pdf (last checked by author April 2002)

Agarwal, B. (1997) 'The gender and environment debate: lessons from India', in N. Visvanathan *et al.* (eds.), *The Women, Gender and Development Reader*, London: Zed Books

Agarwal, B. (2001) 'Participatory exclusions, community forestry, and gender: an analysis for South Asia and a conceptual framework', *World Development* 29(10): 1623-48

Centre for Science and Environment (1999) *State of India's Environment: The Citizen's Fifth Report*, New Delhi: CDE

Chopra, K. and A.K. Duraiappah (2001) 'Operationalizing Capabilities and

Freedom in a Segmented Society: the Role of Institutions', paper presented at a conference on 'Justice and Poverty: Examining Sen's Capability Approach', Cambridge, UK, June 2001, Winnipeg: International Institute for Sustainable Development

FAO (1997) *SD Dimensions: Asia's Women in Agriculture, Environment and Rural Production: India,* http://www.fao.org/sd/WPdirect/WPre0108.htm (last checked by author April 2002).

Government of India (2001) *Census 2001* (provisional), http://www.censusindia.net/results/2001_Census_Data_Release_List.htm (last checked by author April 2002)

IPCC (2001) *Climate Change 2001: Impacts, Adaptation, and Vulnerability*, Geneva: Intergovernmental Panel on Climate Change

Kumar, K. and J. Parikh (1998) 'Climate change impacts on Indian agriculture: the Ricardian approach', in Dinar *et al.*, *Measuring the Impacts of Climate Change on Indian Agriculture*, World Bank Technical Paper 402, Washington DC: World Bank

McKibben, B. (1996) 'The enigma of Kerala', *Utne Reader*, March-April: 103-12

Nussbaum, M. (2000) 'Women and work – the capabilities approach', *The Little Magazine* 1(1), http://www.littlemag.com/martha.htm (last checked by author April 2002)

Patel, V. (2002) 'Of famines and missing women', *Humanscape* 9(4), http://humanscapeindia.net/humanscape/new/april02/culturematters.htm (last checked by author April 2002)

Sanghi, A. (1997) 'Global Warming and Climate Sensitivity: Brazilian and Indian Agriculture', unpublished PhD thesis, Department of Economics, University of Chicago, Chicago IL

Sen, A. (1999) *Development as Freedom*, New York: Anchor Books

Simmons, C. and S. Supri (1997) 'Rural development, employment, and off-farm activities: a study of rural households in Rurka Kalan Development Block, north-west India', *Journal of Rural Studies* 13(3): 305-18

UNDP (1997) *Human Development Report*, New York: UNDP

Venkateswaran, S. (1995) *Environment, Development and the Gender Gap*, New Delhi: Sage Publications

Promoting the role of women in sustainable energy development in Africa:

networking and capacity-building

Tieho Makhabane

The issue of sustainable energy development is a key consideration for climate change mitigation and adaptation initiatives, and is an integral component of Africa's ability to achieve the inter-related economic, social, and environmental aims of sustainable development. Nearly one-third of the global population lacks access to energy-efficient services that do not degrade the ecosystem or contribute to environmental change. Climate change is likely to affect everyone in some way: from rising temperatures, increased floods, and changing rainfall patterns, to the spread of diseases like cholera and malaria (Wamukonya and Skutsch 2001). African countries are likely to be severely affected because of the already high levels of poverty and vulnerability. The impacts of environmental change on men and women are likely to be different with regard to their different roles and responsibilities. This article discusses women's initiatives in the sustainable energy field, highlighting the efforts of two networks that work globally and regionally to strengthen the role of women in sustainable energy development. I highlight some of the challenges that the networks face, and propose strategies for effective networking and capacity-building.

Much of the focus of development interventions in Africa has been on energy-use at community level. For example, many interventions have promoted the use of improved stoves, which end the drudgery of wood-fuel collection by women. The different kinds of impact on men and women of climate change, international energy policies, and climate change mitigation activities, have not been articulated or researched in depth at national and regional levels (Wamukonya and Skutsch 2001). These impacts, and the roles that men and women can play in mitigation activities at local levels, have largely been ignored in international negotiations such as those undertaken through the United Nations Framework Convention on Climate Change (UNFCCC). The gendered implications of climate change and energy policies need to

be explored and addressed, and an understanding of women's particular knowledge and usage of energy resources should be integrated into mitigation initiatives (Wamukonya and Rukato 2001). Networks such as ENERGIA and SAGEN, organising around energy issues, have networking resources, relevant expertise, and an awareness of realities on the ground that can contribute to this task.

In the developing world, 1.3 billion people now live below the poverty threshold, 70 per cent of whom are women (Denton 2001, 4; Misana and Karlsson 2001). Energy-use is a yardstick for socio-economic development, and it is clear that energy poverty and inefficiency are widespread in Africa. Energy-use is closely linked to a range of social issues: poverty alleviation, population growth, urbanisation, and a lack of opportunities for women.

The connections between energy-use and an array of other issues make a focus on energy a key means of achieving greater social justice, including reducing current disparities in power between women and men.

In both developed and developing countries, the challenge ahead requires political will, as well as commitment to innovation, and the application of energy-efficient, environmentally-sound, cost-effective technologies and systems in all sectors of the economy. Energy resources are plentiful, and environmentally sound technological options are available to make a sustainable energy future a reality for all. However, ensuring adequate access to sustainable energy for all will require considerable effort, and substantial investment.

Women's and men's everyday experiences differ in many ways as a result of their differing gender roles and responsibilities. The traditional gender roles of men and women mean that women typically juggle multiple responsibilities in the home, in the workplace, and in the community. As part of their household role, many women are intimately involved in energy-related activities, and have a unique knowledge of the environment and the importance of sustainability. Yet, the demands on women's time and labour, together with widespread social constraints on women's freedom to participate in public action, often leave them with few opportunities for political involvement. This leaves women without a voice in the decision-making processes that affect on their lives and, in particular, their ability to contribute fully to sustainable development.

Developing energy solutions for women's empowerment

Since the energy crisis of the 1970s, there have been considerable advances in addressing the energy-gender gap, as well as attempts to solve the energy bias and meet women's economic needs. Nonetheless, the gender-related challenges within energy provision remain largely unresolved. While energy providers and policy-makers often consider provision of energy as an end in itself, studies have shown this to be of little value, and that the main focus of energy policy should be on the *services* derived from energy. Focusing on service provision means asking whether energy services are accessible, reliable, and affordable, and whether choices and options are available to energy users, in particular the poorest. In Africa, the bulk of policies relating to energy are formulated outside the energy sector, within other development sectors such as agriculture, transport, health, and industry (Denton 2001). These policies consider energy-use as a secondary issue. The result of this type of approach is to ensure that little consideration is given to the energy-provision needs of poor people.

There has been some progress in developing energy solutions for women, and in recognising women's role in sustainable energy development. In Africa, there is a high dependence on traditional fuels. Women are primary users, providers, and managers of energy despite the fact that their access to 'modern' energy sources and technologies is limited (Makhabane 2001). Environmental degradation increases the time that women spend collecting and using household fuels for cooking and heating, and intensifies their workload. Reducing women's workload and the amount of time they spend on it is important for meeting sustainable development goals, and issues around energy are of critical importance within this.

Early development interventions did attempt to move towards sustainable forms of energy-use, taking into account gender-differentiated needs in the developing world. However, the interventions of the 1970s and 1980s concentrated mainly on 'technological fixes', which were seen as the best solution to the energy problems relating to gender and energy concerns (Denton 2001). These 'technological fixes', or supply-side schemes, gave rise to new technologies such as biogas, improved stoves, social forestry schemes, low-grade solar energy systems, wind energy, biomass, and gasifiers. In a bid to assist African women to move towards self-sufficiency in the field of energy, micro-credit schemes and other financial mechanisms with a focus on income-generating activities were introduced.

These early interventions did not tackle the 'real' energy concerns of the users and managers of energy at household level. At a practical level, interventions to provide improved, energy-efficient stoves to women can reduce the risks of indoor air pollution, accidents from open fires, and other related effects on health, as well as reducing time spent by women in gathering firewood. However, the real energy crisis in developing countries, particularly in the rural areas, is related to women's lack of time. Gender-sensitive energy policies and interventions must address women's needs and concerns, taking into account the amount of time that women spend in firewood-collection, food-processing, and collecting water, and not just the provision of improved wood-burning stoves.

Gender-equality activists recognise that women's needs, and their overall socio-economic aspirations, go beyond the provision of improved stoves. Gender-sensitive policies and considerations in the design and implementation of new tech-nologies are crucial. It is argued that if energy provision is to assist women, it must provide technologies and energy services that make women self-sufficient. Self-sufficiency is a critical element in women's empowerment.

International debates and activities focusing on sustainable development have also recognised the role of women, and given significant attention to this.

Concerns over gender and energy problems can be traced back to the United Nation's Third World Conference on Women, held in Nairobi, Kenya, in 1985. Here, women from around the world began to promote their role in sustaining the environment. Two years later, as a result of the lobbying efforts of women, the United Nations commissioned a global study on the environment, from the World Commission on Environment and Development (WEDO 1998). The Commission's team of experts spoke to a broad range of people in all regions about environmental concerns. The team discovered no single priority issue; people identified living conditions, gender issues, lack of resources, population pressure, international trade, education, and health as all being important. As a result, the commission recommended the organisation of an inter-governmental conference, preceded by a participatory discussion process involving civil society. Popularly known as the World Summit, the UN Conference on Environment and Development (UNCED) took place in June 1992 in Rio de Janeiro, Brazil.

For women worldwide, UNCED was an important step in establishing global recognition of their crucial role in achieving a different type of development – a role that is socially, economically, politically, and environmentally sustainable. All UNCED documents included specific recommendations for strengthening women's participation in decision-making processes.

Networking for change

One advantage of networking is that people with the same vision work together to share resources and expertise, while providing each other with support to achieve the desired results and bring about change. In my experience, networking initiatives in which people realise their capacities and capabilities, without underestimating the problems that arise, are most successful when they are underpinned by a shared vision and commitment. In the next section I provide two case studies of initiatives that have attempted to promote the role of women in sustainable energy development.

Case Study 1: ENERGIA – the Global Network on Gender and Sustainable Energy

ENERGIA, the Global Network on Gender and Sustainable Energy, was established in June 1995 by an informal group of women involved in energy inputs to the Fourth United Nations Conference on Women, in Beijing, in 1995. It was established in response to the fact that there was then no international institution or programme with the principal objective of promoting the role of women in sustainable energy development (ENERGIA 2002). The group felt that setting up an international energy organisation to ensure that gender and energy issues were placed on the main-stream agenda of relevant organisations was critical, and that it would be most appropriate coming from the South. With encouragement from the Netherlands Development Co-operation (DGIS), the group established a network to catalyse interest and activities, and a newsletter as a means of communication.

Though global in scope, the ENERGIA network has done most of its initial work in Africa. Having learnt from its experiences in Africa, the network is currently expanding to the rest of the Southern regions. From 1996-8, the central feature of the network was the production of the ENERGIA newsletter. In 1999, a topping-up phase was approved by DGIS, aimed at intensifying involvement of the ENERGIA Support Group through annual meetings, creating a directory of members, and establishing an on-line presence for ENERGIA, including disseminating ENERGIA News (ENERGIA 2002, and see http://www.energia.org/resources/newsletter/index.html).

As a result of numerous requests for other technical services, in 1998 the ENERGIA Consultative Group recommended a number of new and expanded activities for the secretariat and the network. Thus, in July 1999 the ENERGIA programme, with support from the Dutch and Swedish governments, established a network with a permanent secretariat, and a resource centre to support publications and information services. With over 1550 subscribers, two-thirds of whom are based in the South, ENERGIA has undertaken regional networking activities and initiatives in Africa, Asia, and Latin America, and been actively involved in capacity-building activities such as training, visits, and needs-assessments. Research, including the production of case studies on gender and energy issues, and advisory support to gender and energy programmes such as those run by UNDP and the World Bank, have also been important aspects of ENERGIA's work.

ENERGIA's goal is to 'engender energy and empower women', through the promotion of information exchange, research, advocacy, and action aimed at strengthening the role of women in sustainable energy development. The net-work's long-term objective is to promote women's active participation in all areas of energy-use, supply, and management, aiming at redirecting energy policies, research and development, and practice towards the needs of the majority of people, and sustainable development.

Achievements of the ENERGIA Network

ENERGIA has been successful in imple-menting most of its objectives, and has continued to enjoy benefits and support from its members and donors. The following are some of the network's tangible outcomes and achievements:

- The network has created a fully-fledged secretariat, currently hosted by ETC Foundation, the Netherlands, with two

88

directors, a full-time project manager, a full-time project co-ordinator, and two part-time staff.

- The network has established one regional, two sub-regional, and nine national focal points, all of which are in Africa. These national and regional networks adhere to ENERGIA's basic principles, which are to empower women and engender energy for sustainable development.
- The network provides continued support to members of its consultative group as well as to other networks and collaborators.
- The network has achieved consistency in delivering the desired services, as well as in information sharing and exchange.

Case Study 2: The Southern African Gender and Sustainable Energy Network (SAGEN)

The Southern African Gender and Sustainable Energy Network (SAGEN) emanated from the ENERGIA network, after the Gender and Sustainable Development meeting that took place in Nairobi in March 2000. The network subscribes to ENERGIA's principles, and its activities have been co-ordinated with those of the sister network, filtering down to the regional and country levels. With support from the sister network, SAGEN has commissioned and managed three background papers which take a gender perspective on climate change, regional environmental change, rural electrification, and power sector reform.

The network's aims and objectives include (amongst others):

- highlighting the importance and relevance of gender-sensitive policies and practices in Southern Africa;
- paving the way for the increased involvement of women in energy-related processes and projects;

- increasing women's participation in the energy sector, at research, intellectual, and practical levels;
- advocating for the increased inclusion of the issue of income-generation in policies on energy provision, and the need to consider income-generation and energy together;
- sharing information and expertise on energy and gender; and
- collaborating on energy and gender projects.

To date, the network has focused on capacity-building for its members, advocacy for gender mainstreaming, and further research and development work.

Understanding the challenges that networks face

In addition to highlighting, and responding to, the huge energy problems that women face at grassroots level, networks and organisations working in the area of gender and energy have also faced their own internal and external challenges highlighted below.

Short-term versus long-term goals

A major challenge for networking organisations has been the question of how to harmonise their activities with those of policy-makers to enable planned change, and particularly when policy-makers are operating in a crisis-management situation. Research is urgently required, but research takes time, and policy-makers are not always in a position to wait for findings to implement change. Another problem is the fact that the organisations and networks in developed and developing countries have not fully addressed concerns of relevance to specific countries and groups.

Investment and benefit ratios

The sustainability of any project or network depends on the ratio between the benefits that can be derived from such a project and the investment that must be made to realise those benefits. The investment includes both financial expenditure, and the time and resources that partners are willing to put in. It is important for those involved with projects to provide financial contributions for continuity. The benefits, however, should justify the investment. They should be valued and quantified, so that they can be matched to the investment. No project can survive without funding, so either the initiating partners must make a commitment at the beginning of the project to continue funding activities at such a time as the main sponsor pulls out, or the project itself must generate revenue and opportunities from its activities and services.

'Brain-drain' and loss of personnel

Experts and personnel move constantly, and projects often lose valuable personnel. This affects the continuity of projects, and relationships between project members. Training and inductions for new personnel require time and money, and affect the duration of the desired outputs. In some cases, people become part of the project just because the issue that is being dealt with is fashionable and liable to receive funding, without having the necessary long-term vision to make a difference in the field.

Selection criteria of partners

The number of researchers in the field of gender and energy are currently very few, especially in developing countries. The challenge therefore is to move from traditional compositions of energy researchers – predominantly male engineers – to a broad base of researchers in aligned fields, including those with social, economic, and cultural dimensions. Networks also have to create a relationship with policy-makers, for their efforts to be realised. Further, greater efforts must be made to find ways to include project beneficiaries as partners in planning and policy-making. New vision is required to realise this in the field of gender and energy, where local-level participation is still under-researched.

Governance issues

Governance is of overwhelming importance, especially when it comes to networks and project implementation. It involves transparency, reporting, conduct, bureaucracy, structures, implementation mechanisms, and so on. Excessive bureaucracy and corruption are the biggest threats to projects worldwide. Bureaucracy causes delays, escalates costs, and erodes morale. Organisations and institutions are different, but it is important to control governance hurdles so that they do not affect projects negatively.

Unclear objectives and plans

Clear objectives are extremely important when embarking on any project. Plans that are poorly thought-through or unrealistic will not deliver the intended results. Realistic plans will take into account the availability of resources, and will reduce the risk of the project falling behind schedule. It is extremely important that plans are closely monitored, and that reporting takes place frequently. The person who bears the financial burden is the one who is most emphatic about reporting and producing results. This is all the more reason why all partners should make a financial commitment to the project at some point. Project planning at the outset is crucial to ensuring that all obstacles to sustainability are removed. If funding has been released to meet a particular set of objectives that makes the project unsustainable, then this should be clarified at the start of the project so that there is no illusion that the project can be sustainable.

Communication problems can arise

Communication and information-sharing provide fora for a shared vision and common goals. Without these, a network cannot function at all. In most cases, partners rely on the secretariat to initiate and communicate ideas, and there can be a resulting lack of ownership around the implementation of proposed ideas.

Strategies for gender integration into local policies for sustainable development

The service of gender and energy networks will only be of value if their operations reach the beneficiaries. Therefore, networks must identify practical strategies to achieve this.

Networks must find ways and means to advocate for the establishment of 'focal points' in energy departments, with a clear mandate and adequate resources. Governmental focal points for gender and development in most countries have no mandate, and no strategy guidelines for their activities on energy and gender. Many are not given adequate budgetary allocations for their work. Networks must strive to strengthen their contacts and collaboration with these actors.

The importance of energy to sustainable development needs to be widely recognised, with sufficient research and resources directed toward sustainable energy policies and interventions (DFID *et al.* 2001). An understanding of energy as a development issue has only gained prominence in recent years, particularly in relation to rural development initiatives. As stated earlier, most rural development strategies have never brought the issue of energy to the fore. When it has been included, it has been subsumed in sectors like agriculture or transport. Because of this, there has been very little attempt to consider energy issues in their own right. The gender issues in household energy use need more research, including more case studies to examine household energy-use patterns, and to explore how women's unrecognised labour can be incorporated into national development policy. A policy is needed that assists by setting out steps towards co-ordination and strengthening of the complementary roles that NGOs and local institutions can play in the field of gender and energy. More studies are needed in the field of energy into ways of strengthening local women's participation within NGOs and government.

A compilation of gender-disaggregated data could assist in creating activity profiles, including women's household and productive roles in the energy sector, as a basis for lobbying and to facilitate on-going research in gender and energy (Wamukonya and Rukato 2001). This must be made available in a user-friendly form. Networks must strive to reach poor rural women in order to assist in identifying projects, and to encourage local women's initiatives in a drive towards decentralisation of the network and a move away from top-level operations. The networks must have roots that reach the local level, in order to address micro-level needs.

Networks must strive to run gender and energy-related programmes, and provide training to sensitise staff and policy-makers within energy-providing and government institutions. In particular, they need to provide training in applying gender analysis appropriately within energy-related work. This training would help in the implementation of projects that address gender issues. Effort is needed in lobbying for an increase in the number of women working in the energy sector and in policy-making, and ways need to be found to encourage and support more women in this area. This could be through the identification of career development programmes, and the involvement of women in the government and non-governmental sectors.

Energy is not a sectoral issue, and because of this, networks need inter-disciplinary and multi-sectoral staff and teams. Laws and other legal instruments that hamper women's empowerment and development, or that encourage and perpetuate gender discrimination and impinge on women's position and participation, should be amended.

Conclusion

Concerted action to address the connections between gender, energy, poverty, and climate change is possible through the sharing of existing knowledge and resources. We need to create and support a network of organisations working on climate change and energy issues, through the commissioning of background papers to influence national, regional, and international climate change policy-making processes. These organisations need to participate at all levels of political advocacy and policy-formulation, including the forthcoming World Summit on Sustainable Development, to be held in Johannesburg in 2002, and the preceding preparatory committee meetings.

It is particularly important that efforts should consider carefully where and how gender issues need to be addressed in the climate change debate and focus on these, rather than tackling the problem broadside. In particular, it is necessary to develop a strategy that is practical, incorporating general concerns expressed in terms of the need to involve women in decision-making, and the need to respond to women's needs for real opportunities for development.

Tieho Makhabane is a consultant in the field of gender, energy, and development. Contact: Gender and Energy Programme Specialist, Gender, Energy, and Development (GED), P.O. Box 747, Buccleuch, 2066 Sandton, South Africa. Tel: +27 11 656 0601; Fax: +27 11 802 0041; Mobile: +27 73 237 2683; E-mail: tieho@mepc.org.za or ttheoha@yahoo.com

References

Denton, F. (2001) 'Gender and Energy Workshop: Moving Towards Practical Solutions for Meeting Gender Differentiated Energy Needs Within an Integrated Development Approach', Dakar: ENDA Tiers-Monde

DFID, EC, UNDP, and the World Bank (2002) 'Linking Poverty and Environmental Management: Policy Challenges and Opportunities', discussion document for contribution to WSSD, Johannesburg, 2002

ENERGIA (2002) 'Progress Report: July 1999 to December 2001', Leusden, Netherlands: ENERGIA secretariat

Makhabane, T. (2001) 'Gender and Energy in Southern Africa, Draft Proposal developed for the Southern African Gender and Energy Network (SAGEN)', South Africa: MEPC

Misana, S. and G. Karlsson (2001) *Generating Opportunities: Case Studies on Energy and Women*, UNDP: New York

Wamukonya, N. and H. Rukato (2001) 'Climate Change Implications for Southern Africa', background paper prepared for the Southern African Gender and Energy Network, South Africa: MEPC

Wamukonya, N. and M. Skutsch (2001) 'Is there a Gender Angle to Climate Change Negotiations?', New York: Commission for Sustainable Development

WEDO (1998) *Women Transform the Mainstream, 18 Case Studies of Women Activists Challenging Industry, Demanding Clean water and Calling for Gender Equality in Sustainable Development*, New York: Commission for Sustainable Development

Transforming power relationships:
building capacity for ecological security

Mary Jo Larson

The risks that climate change poses for the environment and for development are well-documented, yet it has been difficult to build a consensus on measures to reduce global threats to ecological security. How can communities, NGOs, and policy-makers representing less powerful nations overcome objections to measures that aim to mitigate the global threat to environment and development? In climate change negotiations, vulnerable communities and disadvantaged groups meet around the same table as more powerful interests. Using systems theory,[1] this article analyses the ways in which low-power groups can transform disadvantageous power relations to overcome threats to sustainable development.

In this article, I propose a holistic, integrated, flexible approach to ecological security, which would strengthen the power (symbolic, social, and material) of disadvantaged groups. This would potentially enable them to overcome barriers to healthy, sustainable development. Systems theory suggests that capacity-building can contribute to ecological security by enabling low-power groups to appreciate, influence, and manage their strategic interests. Lessons from analyses of multilateral negotiations are relevant to women, and to any group whose lack of power results in threats to their ability to live and develop sustainably.

The article is structured in three sections. The first describes the risks of climate change to vulnerable small island communities in the Pacific. Island and coastal nations are among the populations most at risk of the effects of climate change.

The second uses a conflict resolution systems approach to investigate the priorities of 43 small island nations from the Pacific, Caribbean, and Indian oceans, in UN climate change negotiations. Situated in a position of low power within the international system, these nations have formed the Alliance of Small Island States (AOSIS) to heighten global awareness of the threats to their ecological security. The section maps the ways in which AOSIS tries to transform threatening and unequal relationships. The third section analyses the small islands' proposals, in order to identify their strategic interests, before adopting a systems approach to discuss the ways in which capacity can be built in order to promote ecological security. The article is based on a combination of academic research on climate change negotiations, and international development experience.

Threats to basic needs in the Pacific islands

Climate change is associated with major health risks and irreversible environmental damage. In the Pacific islands, warming is destroying coral reefs, reducing precipitation, and causing sea levels to rise. Each of these environmental changes has direct and indirect social consequences. Climate change threatens the basic needs of island communities for land, fresh water, food sources, and livelihoods. This in turn affects community health. Poor women, men, and children, who have limited access to land, potable water, cash, or credit, are least able to adapt. The effects of climate change were reported by small island representatives as follows:

Warming trends cause rising sea levels.
>> Rising tide tables cause lands to be inundated.
>> Salt infiltration destroys farm crops.

Warming trends reduce precipitation.
>> Drought dries water reservoirs.
>> Drought destroys root crops.

Warming trends in water temperatures cause extensive coral bleaching.
>> As coral reef communities die, the loss affects marine habitats.
>> Access to coastal fish decreases accordingly.

The patterns above demonstrate the complexity of the links between the environment and development. The decline in agriculture and fishing limits people's access to cash. Reduced access to cash limits the availability of technical goods and services. The socio-economic risks of these changes are exacerbated by poverty, and women are disproportionately represented among the poor and less-educated. They and their dependants are among the most vulnerable community members.

The inter-relationships between environmental habitats and human security extend beyond the issues noted above. Small islands are susceptible to hurricanes, typhoons, floods, and tidal waves. The loss of coastal habitats, such as mangroves and coral reefs, reduces the natural barriers to ocean fluctuations. People, homelands, local cultures, and political independence are at risk. As the Vice-President of Palau has stated, 'This high sea level rise has literally caused islands to disappear and others are in eminent danger of disappearing.' (Remengasau 1999, 4)

Analysing power and relationships

AOSIS faces three major challenges in climate change negotiations. The first challenge has a *symbolic* dimension. This alliance is concerned with the creation of global ecological understandings and commitments. The effort to mobilise the international community in support of sustainable ecological policies rests primarily on the capacity to generate a commitment to shared meanings (Rapoport 1997).

Climate change is an abstract threat to security, and the needs of the 43 coastal and small island states in AOSIS do not have high international visibility. The 1992 Framework Convention on Climate Change does acknowledge that climate change is a human-induced threat, and that the environmental changes adversely affect coastal and small islands nations. However, the powerful parties most responsible for high levels of greenhouse gas emissions have used contradictory studies to dispute the scientific analyses of global environmental changes. Reports purveying mixed messages contribute to public confusion, and have undermined efforts to mitigate the threats through policy-making.

The second major challenge to AOSIS has a *social* dimension. Small island and coastal states are often isolated geographically, and they function on the margins of international policy-making structures.

Despite the fact that 43 nations have enhanced their social power by forming the AOSIS alliance, AOSIS lacks the political mechanisms to regulate the international practices that threaten its constituents' ecological security. Powerful industrialised nations have resisted international efforts to regulate greenhouse gas emissions. They are keen to avoid policies, laws, regulations, or other structures that might reduce the benefits of current practices.

Finally, there are tangible constraints to the security of small island nations. AOSIS does not have the technical or financial resources to mitigate the threats of climate change. While attempting to build ecological security, small islanders face the most powerful corporations on Earth – in particular, those in energy, transport, and agriculture. These powerful groups have differing perspectives on the causes of climate change, the extent of environmental threats to human security, and the most effective responses to undesirable conditions.

To summarise, the parties engaged in climate change negotiations are in conflict, struggling for significant symbolic, social, and material resources (Docherty 1998). As noted above, the effort to reduce fossil fuel emissions is just one dimension of the complex negotiations. Other factors contributing to the conflict and its resolution include differing beliefs and values, and differing approaches to authority.

Conflict is a means of responding to the security threats associated with climate change. 'As a stimulus for the creation and modification of norms, conflict makes the readjustment of relationships to changed conditions possible.' (Coser 1956, 128) Reactions to conflict can be viewed on a continuum, with aggression on one end of the spectrum, and collaboration at the other. Between these extremes are flexible options, including co-operation and competition. Parties prefer differing levels of integration (inclusion and exclusion), depending on the form of power under consideration.

Figure 1: *Levels of flexibility*[2]

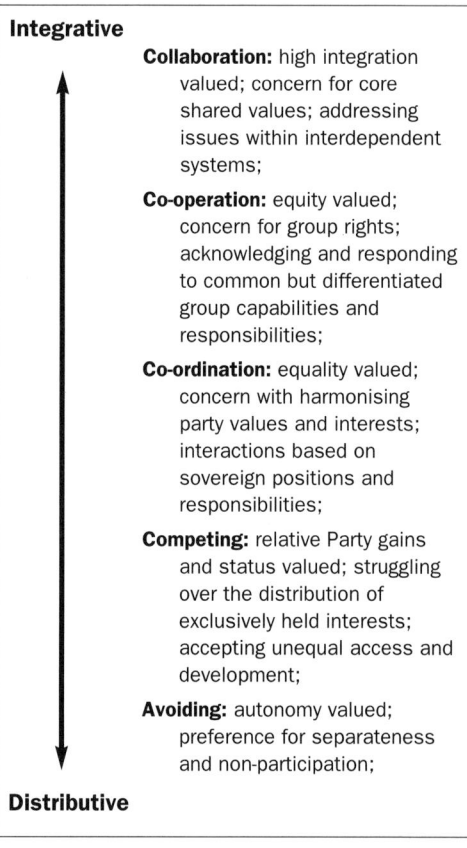

The figure above demonstrates a continuum of options in response to conflicts. Integrative responses, such as collaboration and co-operation, develop common understandings, structures, and interests. An example is the UN's recognition of climate change as a common global concern to humankind. Distributive approaches, such as competition, reinforce autonomy and relative advantages.

Research indicates that flexible (integrative and distributive) behaviour is at the heart of the conflict resolution taking place in negotiations (Druckman and Mitchell 1995). My analysis of the 1992 and 1997 climate change agreements supports the proposal that multilateral negotiations contribute to the resolution of ecological conflicts through functions that are both integrative (collaboration and co-operation) and distributive (co-ordination and competition). Most of the statements in the 1992 Framework Convention on Climate Change and 1997 Protocol are integrative.

Technical experts, particularly NGO representatives, are uniquely situated to facilitate flexible, multilateral approaches to ecological conflicts. Respected facilitators are women and men with access to both policy-makers and stakeholders at the grassroots levels (Lederach 1997). They are most effective when they are able to build relationships across diverse interests, ages, regions, professional groups, genders, cultures, and socio-economic classes. Mediated interventions contribute to conflict resolution when agreements strengthen the most vulnerable groups within the multilateral system (Clements 1993).

Analysing strategic interests

What are the strategic preferences of small island nations as they negotiate to build ecological security? I analysed the 1994 AOSIS Draft Protocol to the UN Framework Convention on Climate Change (AOSIS 1994). Figure 2 shows the conflict resolution systems framework that I chose to use. The conflict resolution systems framework is a tool used to categorise and compare the interests of various parties in negotiations. It can be used to monitor changing preferences over time. The approach recognises that preferences may vary, depending on the substantive issues under consideration.

The nine options in the matrix relate three dimensions of power to three levels of flexibility.

Most of the statements in the 1994 position paper proposing to enhance AOSIS *symbolic* power (28.6 per cent) are at the collaborative level of integration. The 1994 position paper begins with statements of principles, common ecological understandings, and definitions.

'Acknowledging the ultimate objective of the Convention is to achieve stabilization of greenhouse gas concentrations at a level that would prevent dangerous anthropogenic interference with the climate system...' (AOSIS 1994, 3)

Figure 2: *Using the conflict resolution systems framework to analyse the strategic interests of AOSIS*

	Symbolic power	Social power	Material power
Collaboration	28.6% Common ecological understandings	0% Centralised laws with sanctions	1% Core UN funding
Co-operation	9.2% Equitable group rights and responsibilities	44.9% Equitable roles and rules	4.1% Equitable exchanges of technical, financial, and other materials
Co-ordination and competition	1.0% Equal stakeholder rights and relative interests	11.2% Sovereign authority and responsibility	0% Equal benefits and competitive interests

'The Meeting of the Parties shall review and revise the commitments of the Annex I Parties... in accordance with the precautionary principle and the best available scientific information...' [3] (op. cit., 5)

AOSIS is reaffirming meanings established in the 1992 Convention.

A notable percentage of the symbolic statements (9.2 per cent) are co-operative proposals that apply principles of equity to the establishment of rights and responsibilities. Examples of AOSIS' equitable approach include:

'Reaffirming that per capita emissions in developing countries are still relatively low and that the share of global emissions originating in developing countries will grow to meet their social and development needs.' (op. cit., 3)

Most of the statements in AOSIS' 1994 position paper address social power. These are statements prescribing policies, procedures, roles, and dispute resolution mechanisms. AOSIS is attempting to establish predictability and regulate threatening emissions levels. A high percentage (44.9 per cent) of these prescriptions focus on equitable roles and rules. They are co-operative efforts to establish authority and accountability. Examples include:

'Each of the Annex I Parties shall reduce its 1990 level of anthropogenic emissions of carbon dioxide by at least 20 percent by the year 2005.' (op. cit., 4)

'A Meeting of the Parties shall at its first Meeting, agree upon and adopt by consensus, rules of procedure and financial rules for itself and for any subsidiary body.' (op. cit., 7)

The 1994 AOSIS position paper does not emphasise *material* power. Only 4.1 per cent of the statements propose equitable exchanges of resources. They support technology transfer, co-operative approaches to natural resources, and equitable access to economic incentives. The emphasis that

AOSIS gives to equity reflects a strategic interest in maximising technical support to dis-advantaged parties. For example:

'Annex I Parties shall ensure that every practicable step is taken to support the development and enhancement of the endogenous capacities and technologies of developing country Parties.' (op. cit., 7)

It should be noted that AOSIS has not proposed that the UN have sanctioning power. In fact, none of the statements in the 1992 Convention or 1997 Protocol support coercive mechanisms. Instead, through UN climate change negotiations, delegates from AOSIS have formed a global alliance of 43 nations, gained access to vital information, influenced conflict resolution mechanisms, and initiated preventative measures at local levels. Some of their initiatives have been more successful than others. The challenge to AOSIS, and to other vulnerable groups, is to learn from this capacity-building experience.

From research to practice

In this section of the paper, I apply the lessons from the analysis of climate change negotiations to the issue of capacity-building. The success of capacity-building is related to the leadership capacities of the participants. Local stakeholders require the confidence, knowledge, and skills to strategise and communicate effectively with policy-makers, the media, and other opinion leaders. Policy-makers must be capable of understanding and responding to the values, beliefs, and security interests of disadvantaged groups.

Within the context of systems theory, I view capacity-building as a flexible, multilateral approach to sustainable development. Capacity-building is flexible when integrating strengths and respecting different interests. Approaches are multilateral when involving policy-makers, NGOs, and community leaders in self-determined approaches to development. This enhancement of knowledge, skills, and

Figure 3: *Relating the strategic interests of AOSIS to capacity-building*

	Symbolic power	**Social power**	**Material power**
Collaboration	Policies improving common ecological understandings*	Policies and laws with sanctions	Policies managing government and donor funding
Co-operation	Establishing equitable rights and responsibilities*	Developing equitable roles and rules*	Managing equitable exchanges of technical and financial materials*
Co-ordination and competition	Defining equal rights, and sovereign and competitive interests	Establishing local authority and responsibility*	Managing sovereign resources and competitive interests

attitudes is supported through technical resources, including basic tools and sophisticated systems, such as the Internet.

A conflict resolution systems framework provides a way of prioritising the capacity-building interests of parties in multilateral negotiations. It serves as a resource for identifying common ground, acknowledging differences, visualising options, and strategic planning. The framework is able to highlight strengths and barriers within the negotiations. It also provides support for monitoring and evaluation.

Ideally, the different parties in multilateral negotiations can use the systems framework to identify their own capacity-building priorities. For illustrative purposes, I will discuss the strategic interests of AOSIS as they might relate to capacity-building. I believe that these interests are relevant to many disadvantaged groups. In Figure 3, the conflict resolution systems framework serves as a graphic organiser for the development of a systems approach to capacity-building.

What are the priorities of disadvantaged groups as they respond to threats to their environment and development? The five strategic interests of AOSIS are highlighted above with asterisks (*). They are organised under three dimensions of power.

Symbolic power

1. The first strategic interest is to improve common understandings. This symbolic form of empowerment establishes a rationale for action. As noted above, NGO facilitators are able to contribute to ecological security by building respectful commitments among policy-makers and representatives of local stakeholders. The parties should establish common ground early in the consensus-building process. At this level, capacity-building is the inclusive enhancement of awareness, knowledge, respect, and commitments. The emphasis on collaboration is an effort to increase transparent access to knowledge about perceived security threats, scientific research, socio-economic indicators, political events, proposed policies, and technical initiatives.

2. A second strategic priority is the development of equitable rights and responsibilities. The emphasis on principles of equity recognises that parties in complex ecological negotiations have common yet different needs, responsibilities, and capabilities. To address imbalances in status and knowledge, capacity-building provides opportunities for disadvantaged groups to build research, strategic communication, and knowledge-management systems.

It is important that the participants expect and explicitly recognise uncertainty, and have the capacity to cope with ambiguity. 'Courtesy is one of the great human inventions for bridging uncertainty.' (Bateson 1994, 13) The following questions help to elicit the multiple views of situations held by participants, including perceptions of power relationships:

- What are the major concerns or issues? Are women's and men's perspectives equally represented?

- Who are the significant stakeholders? What are their primary interests?

- What are the risks – from diverse perspectives? How are the risks distributed among the parties?

- Who benefits from the current situation? What are the constraints to resolution?

- What are the optional solutions? What are the recommended next steps?

A multilateral approach expands the flow of information from high- and low-power perspectives. Donor communities enhance capacity-building by supporting transparent data collection and analysis, and by disseminating best practice. Accurate and transparent data-collection systems reduce the possibility of wastage, corruption, and inefficiency. The challenge is to assure that dialogue leads from information exchange to strategic adaptations, including opportunities for women and men to build their own data-collection systems within their own jurisdictions.

Social power

3. The majority of the propositions in multilateral negotiations are prescriptive. Disadvantaged groups are attempting to address their strategic interests through the development of equitable roles and rules. The emphasis on equity in policy-making has a re-distributive purpose. The strategic interest is to create more balanced power relationships. Those most responsible for threats should take immediate steps to mitigate the risks.

The challenge to disadvantaged groups is to transform the structures that support existing power imbalances. When high- and low-power groups are engaged in policy-making, the advantaged parties tend to defend their territory and status. Inclusive, transparent approaches to rule-making enhance the leverage of disadvantaged groups. By influencing policies and laws, they are able to improve access, predictability, and accountability. Questions relevant to the transformation of social structures include the following:

- Who has authority? Under what circumstances?

- Do women and men participate equally in the decision-making?

- Which policies and laws contribute to fairness and predictability?

- Which governance structures are missing?

- Do laws foster equitable (fair) multilateral partnerships?

- How can threats to security be regulated more effectively?

It should be noted that disadvantaged groups are not advocates for centralised power with sanctions. Instead, most of the proposals to regulate threatening behaviours involve consensual agreements. Strategic priorities include developing the capacities required to mobilise coalitions.

4. The fourth strategic interest is the establishment of sovereign authority and responsibility. This priority addresses the need to build effective local governance structures. It suggests that all parties have equal responsibility to mitigate ecological risks within their own jurisdictions. Capacity-building initiatives enable

disadvantaged groups when they strengthen participatory approaches to governance. Policies that are developed by women and men closer to the realities on the ground are more effective than those that are based on distant versions of local possibilities and preferences. By co-ordinating national and local exchanges of information about best practices early in the policy-making process, capacity-building reinforces effective structures and avoids unnecessary conflicts.

Material power

5. Finally, disadvantaged groups have a strategic interest in the equitable exchange of material resources. This requires the capacity to manage technical, financial, and human resources. Historically, the transfer of technologies has been criticised for being donor-oriented. There has been a tendency to supply local communities with whatever technologies are available, rather than what is actually wanted or needed. Investments in technical resources address the needs of the disadvantaged when they facilitate partnerships and foster self-reliance. Questions relevant to effective, efficient technology transfer include:

- Which local technologies are working? What adaptations are needed?
- Are creative local initiatives encouraged? What are the incentives?
- What are the benefits of new financial and technical interventions?
- How are the benefits of technical and/or financial interventions distributed?
- Who gains? Who loses?
- Do women and men participate equally in planning and managing? Do they also share equally in the benefits?

At the local level, the priority is to increase women's and men's capacities to adapt technologies to their changing environments. In order to build sustainable systems, they need the attitudes, knowledge, and skills required for assessment, implementation, monitoring, trouble-shooting, and creative innovations. In order to exchange goods and services, developing economies also need access to the consumer markets of wealthier nations.

Lessons learned

The analysis of small island experiences in climate change negotiations provides lessons for any vulnerable or marginalised group determined to re-align power relationships. Existing power asymmetries may be the result of discrimination on the basis of gender, race, religion, or some other categorisation. To summarise, the analysis above indicates that disadvantaged parties addressing ecological security negotiate to enhance five strategic interests:

- common ecological understandings;
- equitable rights and responsibilities;
- equitable roles and rules;
- sovereign local authority and responsibility;
- equitable exchanges of technical and financial materials.

The first strategic interest is the development of common understandings. This symbolic form of empowerment is addressed through collaborative approaches to research and communication. The establishment of common meanings provides a rationale for action. Moving from rhetoric to action, the transformation of power relations takes place through the equitable exchange of resources.

Disadvantaged groups are advocates for equitable access to scientific knowledge, the consensual construction of rules regulating threats, and the equitable transfer of technical resources from developed to developing nations. This co-operation transforms power relations when disadvantaged women and men are able to make choices that enhance the resources within their own jurisdictions. The strategic interests above are relevant to gender in a wide range of development situations, including efforts to address complex environmental and health issues.

Conclusions

In climate change negotiations, small island nations are contributing to the resolution of ecological conflicts through proactive, co-operative approaches. To transform threatening systems, leaders are building regional alliances, developing extensive communication networks, and advocating as one voice for the security interests of the Earth's ecological system as a whole.[4] Identifying the strategic interests of small island communities provides lessons for the transformation of power relationships, including those between women and men. Disadvantaged groups in climate change negotiations lack status, influence, and control. These factors mirror gender-based barriers to health and development.

A systems approach to capacity-building ensures that disadvantaged groups advance through self-determined approaches to development (adapted from International Women's Conference 2000). It is an adaptive learning and relationship-building process. The most effective interventions value and integrate the strengths of diverse social groups. They facilitate the transfer of knowledge and foster the implementation of inclusive policies. Systems theory supports the proposition that capacity-building contributes to sustainable development when it strengthens the most vulnerable sub-systems within the global ecological system as a whole.

Conflict resolution research indicates that the first step in building sustainable multilateral relationships is to develop common understandings. In closing, I would like to highlight the significance of this.

'If we can find ways of responding as individuals to multiple patterns of meaning, enriching rather than displacing those traditional to any one group, this can make a momentous difference to the well-being of individuals and the fate of the earth. What would it be like to have not only color vision, but culture vision, the ability to see the multiple worlds of others?' (Bateson 1994, 53)

A systems approach brings together diverse policy-makers, NGOs, and community leaders to enhance the 'culture vision' of all stakeholders. The multilateral dialogue offers less-powerful groups the opportunity to describe their own situations, prioritise strategic interests, and implement meaningful adaptations. The most effective capacity-building initiatives build long-term partnerships among advantaged and disadvantaged stake-holders. These relationships facilitate the equitable exchange of resources, and create the synergy for innovative adaptations.

Mary Jo Larson completed her doctoral thesis at the Institute for Conflict Analysis and Resolution, George Mason University. She is currently Director of Capacity Building at CEDPA. Contact: 6427 Cavalier Corridor, Falls Church, VA 22044, USA. E-mail: symmetryintl@earthlink.net

Notes

1 Systems theory is the holistic study of how systems and their sub-systems are organised, how they adapt to changing situations, and how the interests of the sub-systems fit or conflict with those of the whole. According to this theory, a sub-system is a set of inter-related elements, each of which is connected directly or indirectly to every other element, and often with extreme sensitivity. Localised causes within sub-systems may have effects within the system as a whole. I should emphasise that *no system can be known completely* (Richardson 1998). Any study of

complex, multi-party negotiations brings certain variables to the forefront, and may overlook others. Significant factors, such as cultural influences or structural manipulation, may not be documented in this analysis (Avruch and Black 1996).

2 I am grateful to Patrick Triano for helping me to visualise these relationships.

3 The Annex I Parties include the 24 original OECD members, the European Union, and 14 countries with economies in transition.

4 Interviews (1997-2001) with UN Ambassador Slade from Samoa, Chair of AOSIS.

References

AOSIS (1994) 'AOSIS Proposal for a Protocol on the Reduction of Greenhouse Gases', paper presented at Inter-governmental Negotiating Committee (INC) 10, Geneva

Avruch, K. and P.W. Black (1996) 'ADR, Palau and the contribution of anthropology', in A.W. Wolfe and H. Yang (eds.), *Anthropological Contributions to Conflict Resolution*, Athens: University of Georgia Press

Bateson, M.C. (1994) *Peripheral Visions*, New York: Harper Collins

Clements, K. (ed.) (1993) *Peace and Security in the Asia Pacific Region: Post-Cold War Problems and Prospects*, Japan: United Nations University Press

Coser, L.A. (1956) *The Functions of Social Conflict*, New York: Free Press

Docherty, J.S. (1998) *When the Parties Bring Their Gods to the Table: Learning Lessons from Waco*, published doctoral thesis, Virginia: Institute for Conflict Analysis and Resolution, George Mason University

Druckman, D. and C. Mitchell (eds.) (1995) *Flexibility in International Negotiation and Mediation*, London: Sage

International Women's Conference (2000) 'International Women's Conference to Redefine Security', proposal developed by Women's Caucus for Gender Justice, June 22–25 2000, Naja, Okinawa, Japan

Larson, M.J. (2001) 'Conflict Resolution in Ecological Negotiations: How Multi-lateral Negotiations Contributes to the Resolution of Environment and Development Conflicts', unpublished doctoral thesis, Virginia: Institute for Conflict Analysis and Resolution, George Mason University

Lederach, J.P. (1997) *Building Peace: Sustainable Reconciliation in Divided Societies*, Washington DC: United States Institute of Peace

Rapoport, A. (1997) 'Conceptions of World Order: Building Peace in the Third Millennium', paper presented at the Tenth Annual Vernon M. and Minnie I. Lynch Lecture, Institute for Conflict Analysis and Resolution, George Mason University, Virginia

Remengasau, T. (1999) 'Remarks by Vice President of Palau on the Opening of the 54th Session of the General Assembly of the United Nations', Washington DC: Embassy of Palau

Richardson, K.A. (1998) 'Towards an Analytical Methodology for Considering Complex, Poorly Defined Problems', paper for Defence and Evaluation Research Agency, Salisbury, UK

Resources

Compiled by Ruth Evans

Introduction

As many of the articles here note, to date, the literature discussing climate change from a gender perspective is rather scarce. Resources on this subject are mainly web-based, and unpublished. There are, however, more resources and publications available on aspects of gender and disasters, and on gender, the environment, and sustainable development. These have been included here, alongside general resources about the effects of climate change, or related to organisations working on these issues.

Publications

Responding to Global Warming: The Technology, Economics and Politics of Sustainable Energy (1994), Peter Read, Zed Books, 7 Cynthia St., London N1 9JF, UK

Taking a multi-disciplinary policy perspective that integrates engineering, economics, and decision theory, the author proposes an innovative strategy in global efforts to limit climate change, linking energy and forestry, North and South. Read argues that the problem of global warming can be tackled much more affordably than commonly realised, and in ways likely to provide incentives to energy corporations, and to improve the development prospects of many countries in the South.

Survival Emissions: A Perspective from the South on Global Climate Change Negotiations (1999), Mark J. Mwandosya, Centre for Energy, Environment, Science and Technology (CEEST-2000), Dar es Salaam, Tanzania

In this book, Mwandosya shares his perceptions and insights on the climate change negotiations, based on his experience as chair of the Group of 77, and speaker for the Group and China during the climate change negotiations in Bonn and Kyoto in 1997. The book gives background information on the negotiations, as well as the author's analysis and understanding of the negotiations from a Southern perspective. In particular, it discusses the strength of the unity of the Group in linking climate change negotiations with the development agenda.

Global Climate Change: The International Response (1996), Richard E. Benedick, Discussion Paper 19, October 1996, London School of Economics, Centre for the Study of Global Governance, Houghton Street, London WC2A 2AE, UK

This discussion paper outlines the development of the international response to climate change, from the establishment of the Intergovernmental Panel on Climate Change in 1988, and the UN Framework Convention on Climate Change, signed in

Rio in 1992. The paper examines in more detail the options for reducing greenhouse gas emissions, discusses North–South tensions, and outlines key factors for the post-2000 phase.

Fair Weather? Equity Concerns in Climate Change (1999), Ferenc L. Toth (ed.), Earthscan Publications, 120 Pentonville Road, London N1 9JN, UK
http://www.earthscan.co.uk

Taking a cross-disciplinary assessment of fairness and equity issues in the context of global climate change, this book explores the policy dimensions and analytical needs of the negotiation process. Contributors debate a range of equity issues in the global climate change negotiations, such as: how should responsibility for adapting to climate change be distributed? Who should bear the costs of mitigating its impacts, and how should these costs be measured? Their responses to these questions differ, often varying according to the vulnerability, wealth, and level of industrial development of the country in question.

The Way Forward: Beyond Agenda 21 (1997), Felix Dodds (ed.), Earthscan Publications

This book outlines the successes and failures of the first five years following the Earth Summit in Rio de Janeiro in 1992. Drawing on the experience of a range of experts, it provides an analysis of the agreements that were reached, and the stakeholders who are charged with implementing them. It reviews the progress made so far at the inter-governmental, national, and grassroots levels, and offers a summary of the major issues that need to be addressed in the future.

Implementing Agenda 21: NGO Experiences from Around the World (1997), Leyla Alvanak and Adrienne Cruz (eds.), United Nations Non-Governmental Liason Service (NGLS), Palais des Nations, CH-1211 Geneva 10, Switzerland, and Room 6015, 866 UN Plaza,

New York, NY 10017, USA

This collection of contributions from NGOs around the world highlights dimensions of Agenda 21 implementation at the local level that might not otherwise be captured by the international dialogue. In their articles, contributors describe NGO projects and other activities focused on the implementation of the 1992 UN Conference on Environment and Development (UNCED), and discuss how UNCED's new approach to sustainable development affected thinking, programmes, and strategies.

Coping with Changing Environments: Social Dimensions of Endangered Ecosystems in the Developing World (1999), Beate Lohnert and Helmut Geist (eds.), Ashgate Publishing Ltd., Gower House, Croft Road, Aldershot, Hants., GU11 3HR, UK
http://www.ashgate.com

This collection of articles takes a multi-disciplinary approach to the social dimensions of global environmental change. Drawing on regional case studies from many developing countries, the collection explores vulnerability, coping strategies, and societal responses to drought hazards, changing land use, and deforestation, amongst other environmental changes. The editors offer readers a comparative perspective on global environmental change.

The Climate Change Negotiations: Berlin and Beyond (1995), Ian H. Rowlands, Discussion Paper 17, July 1995, London School of Economics, Centre for the Study of Global Governance, Houghton Street, London WC2A 2AE, UK

This discussion paper provides an overview and analysis of the First Conference of the Parties to the Framework Convention on Climate Change (FCCC) held in Berlin in 1995. It pays attention to the specific outcomes of the conference, and analyses more general emerging trends.

Climate Change and Human Health (1996), A.J. McMichael, A. Haines, R. Slooff, and S. Kovats (eds.), assessment prepared by a Task Group on behalf of the World Health Organisation (WHO), The World Meteorological Organisation (WMO), and the United Nations Environment Programme (UNEP), available from WHO, CH-1211 Geneva 27, Switzerland

This assessment study, addressing the health implications of climate change, was initiated after consultations took place in 1993 between representatives of the WHO, WMO, UNEP, International Panel on Climate Change (IPCC), and United States Environmental Protection Agency (USEPA). The consultations revealed an urgent need for a comprehensive study, based on IPCC's newer scenarios and predictions. The study examines the various possible impacts of climate change and stratospheric ozone depletion upon human health, ranging from summertime heat stress, increased production of air pollutants, vector-borne diseases, water-borne and food-borne infections, agricultural productivity, extreme weather hazards, sea level rise, and exposure to ground-level ultraviolet radiation. Finally, the implications of global climate change for research, monitoring, and social-policy response are explored.

CIAT in Perspective 2000-2001: Getting the Better of Global Change (2001), Gerry Toomey and Nathan Russell, International Center for Tropical Agriculture (CIAT)
http://www.ciat.cgiar.org

This issue of the International Center for Tropical Agriculture research newsletter, *CIAT in Perspective*, focuses on global environmental change. It includes an article entitled, 'Risky farming in a hotter world', on a new method devised by scientists for predicting how global climate change will affect tropical farming fifty years from now.

Climate Change and World Agriculture (1990), Martin Parry, Earthscan Publications

In this book, Parry analyses the sensitivity of the world food system, and examines the variety of ways in which it will be affected if climatic changes occur in line with most scientific predictions. After describing the effects on agriculture, estimating the impacts on plant and animal growth, and examining the geographical limits to different types of farming, the author considers a range of possible approaches for agriculture to adapt and so mitigate the impacts of climate change.

The Potential Socio-Economic Effects of Climate Change: A Summary of Three Regional Assessments (1991), M.L. Parry, A.R. Magalhaes, and N. Huu Ninh (eds.), United Nations Environmental Programme, PO Box 30552, Nairobi, Kenya
http://www.unep.org

This report summarises the major conclusions of three regional studies (in Brazil, in Indonesia, Malaysia, and Thailand, and in Vietnam) of the potential impact of climate change undertaken by national governments with the support of the United Nations Environment Programme.

Footprints and Milestones: Population and Environmental Change (2001) The State of World Population 2001, United Nations Population Fund, 220 East 42nd Street, New York, NY 10017, USA
http://www.unfpa.org

This UN Population Fund report includes coverage of environmental trends, with regard to water, food, climate change, forests, habitat and biodiversity, poverty and the environment, women and the environment, health and the environment, and action for sustainable and equitable development. It also includes an appendix of global agreements on human rights, environment and development, reproductive health, and gender equality.

Climate of Hope: New Strategies for Stabilizing the World's Atmosphere (1996), Christopher Flavin and Odil Tunali, World Watch Paper 130, June 1996, World Watch Institute, Washington DC, USA

This paper discusses the growing evidence of climate change, and examines approaches to reducing greenhouse gas emissions and stabilising the climate.

Earth Summit 2002: A New Deal (2000), Felix Dodds (ed.), United Nations Environment and Development and Earthscan Publications

As preparations for Earth Summit 2002 proceed, this book provides a progress-report and agenda for Earth Summit 2002 and beyond. Experts from around the world present an assessment of progress to date, set goals, and examine the mechanisms that will enable the international community to complete the tasks set in Rio, and prepare for future challenges.

Climate Change Co-operation in Southern Africa (1998), I.H. Rowlands (ed.), UNEP and Earthscan Publications

This book shows how co-ordinated action among neighbouring countries could reduce greenhouse gas emissions in ways that are environmentally, economically, and socially beneficial. It presents a framework for analysing regional mitigation options among developing countries, and examines particular proposals for Southern Africa.

'Participatory exclusions, community forestry and gender: an analysis for South Asia and a conceptual framework' (2001), B. Agarwal, *World Development* 29(10): 1623-48

Based on extensive fieldwork among community forestry groups in India and Nepal, and using existing case studies, this article demonstrates how seemingly participatory institutions can exclude

women and other marginalised groups. Agarwal provides a typology of participation, outlines the gender equity and efficiency implications of such exclusions, and analyses factors underlying exclusions. A conceptual framework is developed to help analyse the process of gender exclusion and how it might be alleviated.

'Gender and the environment: traps and opportunities' (1992), M. Leach, *Development in Practice* 2(1): 12-22

This article highlights the dangers of essentialising women's roles and relationship with the environment, and argues for an alternative approach examining dynamic gender-differentiated activities, rights, and responsibilities in the process of natural resource management. Drawing on a case study from Gola forest, Sierra Leone, Leach demonstrates how this approach can help to ensure sustainability and equity in the design of projects concerned with the environment.

Engendering the Environment? Gender in the World Bank's Environmental Policies (2000), P.A. Kurian, Ashgate Publishing Ltd.

This book uses feminist theory and concepts to understand the gendered nature of environmental policy and environmental policy analysis. Based on research on the World Bank's Narmada Dam project in India, this gender analysis of the World Bank's policies offers a critical interrogation of the practice of Environmental Impact Assessment, and argues for better understanding of the process in which gender, class, and culture interact to influence environmental policy-making.

Global Environmental Outlook: UNEP's Millennium Report on the Environment (1999), UNEP, Earthscan Publications

The Global Environment Outlook (GEO) Project was launched by UNEP in 1995, in response to the need for comprehensive, integrated, policy-relevant assessments of the global environment. This extensive report includes background information on the GEO Project, global perspectives, regional analyses of the state of the environment, regional policy responses, future outlook, and recommendations.

'Climate change, gender and poverty – academic babble or realpolitik?' (2001), Fatma Denton, *Bulletin Africain – Point de Vue*, No. 14, ENDA-TM, available on-line at: http://www.enda.sn/Bulletin Africain/010 Fatma DENTON.pdf

This short article by Fatma Denton in the Environment and Development Action in the Third World (ENDA-TM) newsletter, *Bulletin Africain*, addresses the question, 'What has gender got to do with climate change?' Denton comments on the predominately male agenda, and women's lack of participation in policy formulation and decision-making on environmental issues, such as conservation, protection, rehabilitation, and management of the environment.

Environment, Development and the Gender Gap (1995), Sandhya Venkateswaran, Sage Publications India, M-32 Block Market, Greater Kailesh-1, New Delhi 110 048, India, and 6 Bonhill St., London EC2A 4PU, UK

In this comprehensive study, Venkateswaran discusses women's roles in activities relating to the environment, the differential impact of environmental degradation on diverse groups of women, and their almost complete marginalisation from policies and programmes that seek to manage the environment. Drawing on case studies and empirical data from government and NGO development programmes in India, a range of issues are explored, including those related to croplands, common lands, forest and water resources, domestic energy, social forestry, technology, the urban environment, and pollution.

Women and the Environment (1994), A. Rodda (ed.), Zed Books Ltd.

This practical handbook focuses on women's roles as users, producers, and managers of the earth's resources, and shows how environmental degradation affects women's health and basic needs. It demonstrates how women can be a major force for environmental change, particularly through their important roles as educators and communicators, and it highlights the varied ways in which women are involved in the implementation of environmental projects. Includes a glossary of environmental terms, a guide to education and action, and a bibliography and resource guide.

Women, the Environment and Sustainable Development (1994), R. Braidotti *et al.*, Zed Books Ltd.

This book examines alternative visions of development, including 'women, environment, and development'(WED), and ecofeminism, aiming to disentangle the various positions put forward by major actors, and to clarify the political and theoretical issues at stake in the debates on women, the environment, and sustainable development.

Feminist Perspectives on Sustainable Development (1994), Wendy Harcourt (ed.), Zed Books Ltd.

In this book, researchers, activists, and policy-makers from the North and South propose different ways of challenging dominating knowledge systems and development institutions. The contributors discuss themes such as situating the feminist position in the sustainable development debate; gendered alternatives to dominant knowledge systems; politics and resistance in the sustainable development debate; and population.

Environmental Policies and NGO Influence (2001), A. Thomas, S. Carr, and D. Humphreys, Routledge, 11 New Fetter Lane, London EC4P 4EE, UK

This book examines why NGOs are at times able to exert influence on policies to conserve and use natural resources sustainably in sub-Saharan Africa. After developing a conceptual framework and exploring land resource issues in sub-Saharan Africa, the authors examine case studies of NGO activity and conclude with a summary of lessons to be learnt from studies of NGO campaigners and policy specialists.

Sustainable Development and Integrated Appraisal in a Developing World (2000), N. Lee and C. Kirkpatrick (eds.), Edward Elgar Publishing Ltd., Glensanda House, Montpellier Parade, Cheltenham, GL50 1UA, UK and 136 West St., Suite 202, Northampton, Massachusetts 01060, USA

An international group of authors from a range of disciplinary backgrounds present alternative perspectives and methods for an integrated approach to sustainable development. They apply integrated appraisal to a variety of case studies from developing and transitional countries.

Electronic resources

Gender Perspectives for Earth Summit 2002, international workshop, available on-line at: http://www.earthsummit2002.org/workshop

This workshop provided an overview of, and developed recommendations on, gender perspectives from developing and developed countries on energy, transport, and environmental decision-making issues, addressed during the UN Commission on Sustainable Development (CSD) Ninth Session in April 2001. The website provides reports, background papers, and resources related to the three workshop issues.

Climate Change Information Kit, UNEP/IUC, Geneva Executive Center, CP 356, 1219 Châtelaine, Switzerland.
E-mail: iuc@unep.ch
Available on-line at: http://www.undp.org/seed/eap/Publications/2001/2001a.html

This information kit, published by the UN Environment Programme Information Unit for Conventions, provides a series of papers introducing the impacts of climate change, the Climate Change Convention, limiting greenhouse gas emissions, and useful facts and figures, updated in 1999.

The Stakeholder Toolkit for Women, M. Hemmati and K. Seliger (eds.), UNED Forum, 3 Whitehall Court, London SW1A 2EL, UK.
E-mail: info@earthsummit2002.org
Available on-line at:
http://www.earthsummit2002.org/toolkits/women/index.htm

This toolkit is intended to help monitor the progress made in implementing the Global Plans for Action, which have been agreed at the UN Global Summits and Conferences since the Earth Summit in Rio in 1992. This UNED Forum initiative is aimed at women's groups and NGOs working to implement the global agreements and use them in their work, as well as those who contribute to policy-making at local, national, or international levels. The website contains UN documents from the cycle of World Conferences, NGO position papers, examples of good practice, training materials, campaign information, networking, and useful links.

Just a Lot of Hot Air? A Close Look at the Climate Change Convention (2000), The Panos Institute, 9 White Lion Street, London N1 9PD, UK.
Available on-line at:
http://www.panos.org.uk

This briefing provides an accessible introduction to the Climate Change Convention, appealing to a wide range of audiences. It includes research carried out in the UK as part of the Global

Environmental Change Programme of the UK's Economic and Social Research Council.

Environmental Management and the Mitigation of Natural Disasters: A Gender Perspective (2001), UN DAW (United Nations Division for the Advancement of Women), Report of the Expert Group Meeting, Ankara, Turkey, 6-9 November 2001. Available on-line at:
http://www.un.org/womenwatch/daw/csw/env_manage/index.html

This report discusses the linkages between gender, environmental management, natural disaster reduction, risk management, and the role of different actors. It adopts a number of recommendations on policies, legislation, participation, information, capacity-building, research, and the role of the international community.

Generating Opportunities: Case Studies on Energy and Women (2001), Gail V. Karlsson (ed.), United Nations Development Programme (UNDP). Available on-line at:
http://www.undp.org/seed/eap/Publications/2001/2001a.html

This book of case studies was prepared as part of a UNDP project entitled 'Energy and Women: Generating Opportunities for Development', initiated in February 1999 with support from the Swedish International Development Co-operation Agency and the UNDP's Sustainable Energy Global Programme. The publication looks at critical policy and programme design options to improve women's access to modern energy services based on the lessons learned in the eight case studies presented.

Gender, Environment and Development Guide, Irene Dankelman, UNIFEM.
Available on-line at:
gopher://gopher.undp.org/11/unifem/poli-eco/eco/susta/ged

This guide, aimed at UNIFEM staff and consultants, provides a framework for the evaluation of projects and programmes from an environmental perspective. An overview of how environmental issues impact on UNIFEM-supported projects focuses on how women gather and process natural resources, the environmental impact this has, and how to do a cost-benefit analysis of these activities, incorporating economic, ecological, and social perspectives. Organisational tools for gathering information about the gender, environment, and development (GED) aspects of projects are presented, including checklists of questions, and sample projects with recommendations to improve environmental aspects of the project. Models for analysing information gathered from the checklists are provided, and measures for mitigating negative environmental effects are explored. Environmental and social indicators for monitoring project developments are also identified.

Gender, Conservation and Community Participation: The Case of the Jaú National Park, Brazil (1999), R. Oliveira and S. Anderson, Managing Ecosystems and Resources with Gender Emphasis (MERGE), Case Study No. 2, Center for Latin American Studies, University of Florida. Available on-line in English, Portuguese, and Spanish at:
http://www.latam.ufl.edu/publications/publisting.html

The Fundação Vitória Amazônica (FVA) is a local NGO which has carried out pioneer work on gender, community participation, and partnership building in their conservation activities in the Jaú National Park (PNJ). They are part of the MERGE (Managing Resources and Ecosystems with

Gender Emphasis) programme, funded by USAID and co-ordinated by the University of Florida. PNJ is the largest National Park in Brazil, and the largest protected area of tropical forest in the world. During the consultation process, participatory and gender-sensitive approaches such as informal interviews, gender analysis, and gender mapping of natural resource use, were used to measure men's and women's use of natural resources in subsistence fishing, hunting, and agriculture and their commercial use of natural resources such as vines and Brazil nuts. The findings helped FVA to re-evaluate and adjust their work to involve the community in resource management.

Development and Gender in Brief (1995), BRIDGE, Institute of Development Studies, University of Sussex, Brighton BN1 9RE, UK. Available on-line at:
http://www.ids.ac.uk/bridge/dgb1.html
This issue of 'Development and Gender in Brief' asks whether recent changes in environmental policy have produced real benefits for women, and presents evidence that suggests that many projects fail to promote women's interests. It highlights how, for example, attempts to address the wood shortage in Ghana have been biased towards men, and how special efforts are required to extend women's participation in water and sanitation activities beyond their traditional roles. The same is true of responses to environmental disasters such as the 1991 Bangladesh cyclone; women's needs were neglected leading to higher mortality among women than men.

Reaching the Goals in the S-21: Gender Equality and Environment, Vol. III (1999), Swedish International Development Co-operation Agency (SIDA), OECD Development Assistance Committee, Working Party on Gender Equality. Available on-line at:
http://www.oecd.org/dac/Gender/pdf/wid993e.pdf

Intended to support agencies' implementation of the OECD Development Assistance Committee's policy statement, 'Shaping the 21st Century: the Contribution of Development Co-operation' (1996), this report presents the key findings and good practice from the reports of seven bilateral agencies. These efforts are discussed in relation to: policy; institutional/organisational level; policy dialogue; tools and methodologies; monitoring and evaluation; and donor agency capacity. Five key areas for future attention are discussed: (1) getting agency fundamentals right, particularly in areas of leadership, allocation of responsibility to all staff, and dedication of resources; (2) documenting the key linkages/rationales for the consideration of gender equality and environmental sustainability; (3) moving the analysis and focus up from the community level to include broader spheres of activity, such as gender issues in institutions involved in decision-making around environmental issues; (4) bringing a gender-equality perspective to capacity development in environment; and (5) moving towards mainstreaming strategies with an 'agenda-setting' focus.

Pro-Poor, Gender- and Environment-Sensitive Budgets Project (1999), United Nations Development Programme (UNDP). Available on-line at:
http://www.undp.org/poverty/initiatives/budgets.htm
This UNDP project examines various initiatives from different countries which have attempted to review the impacts of national development strategies and budgets on structural injustices such as gender inequality, poverty, and environmental degradation. Drawing on lessons learned from such international initiatives, the project identifies appropriate future strategies and synergies with other policy areas, with the objective of reorienting macroeconomic policies to meet the needs

of marginalised groups. Activities include co-hosting of national and regional workshops with the United Nations Fund for Women (UNIFEM), input into international conferences, publication of working papers, development of a budget resource book, and an interactive web space for budget-related resources and discussion groups.

Journals

Women and Environments International Magazine, associated with the Institute for Women's Studies and Gender Studies, New College, University of Toronto. Available on-line at: http://www.weimag.com/

Women and Environments International Magazine is an on-line Canadian journal that examines women's relationships to their environments – natural, physical, built, and social – from feminist perspectives. It provides a forum for academic research and theory, as well as professional practice and community experience.

Organisations

Grassroots Organisations Operating Together in Sisterhood (GROOTS) International

Communications, 249 Manhattan Avenue, Brooklyn, NY 11211, USA. Tel: +1 718 388 8915; Fax: +1 718 388 0285; E-mail: grootsss@aol.com http://www.groots.org

GROOTS operates as a flexible network linking leaders and groups in poor rural and urban areas in the South and North. The network is open to grassroots groups and their partners who share a commitment to: strengthening women's participation in the development of communities; helping urban and rural women's groups identify and share successful development approaches and methods; focusing international attention on women's needs and capabilities; and increasing the opportunities

for local women's groups and leaders to network directly across national boundaries.

Intergovernmental Panel on Climate Change (IPCC), http://www.ipcc.ch/

The Intergovernmental Panel on Climate Change (IPCC) was established by the United Nations Environmental Programme (UNEP) and the World Meteorological Organization (WMO) in 1988 to assess the scientific, technical, and socio-economic information relevant for the understanding of human-induced climate change, its potential impacts, and options for mitigation and adaptation. The IPCC publications include assessment reports of the three working groups on the science, impacts, adaptation and vulnerability to, and mitigation of climate change, as well as guidelines and methodologies, special reports, and technical papers.

Foundation for International Environmental Law and Development (FIELD), 52-53 Russell Square, London WC1B 4HP, UK. Tel: +44 (0)20 7637 7950; Fax: +44 (0)20 7637 7951; E-mail: field@field.org.uk http://www.field.org.uk/

The Foundation for International Environmental Law and Development (FIELD) was founded in 1989 in order to tap the potential of law at the international, regional, and domestic levels, and to encourage environmental protection and sustainable development. FIELD provides advice to governments, non-governmental organisations, inter-governmental organisations, and industry.

Pew Center on Global Climate Change, 2101 Wilson Blvd., Suite 550, Arlington, VA 22201, USA. Tel: +1 703 516 4146; Fax: +1 703 841 1422 http://www.pewclimate.org/

The Pew Center on Global Climate Change is a non-profit, non-partisan, and independent organisation that aims to educate the public and key policy-makers about the causes and potential consequences of climate change, and to encourage the domestic and international community to reduce emissions of greenhouse gases. Its activities include releasing reports on environmental impacts, economics, and policy issues; educating the public through advertising, public-speaking events, and conferences; and co-ordinating policy, industry, and government discussions to advance international negotiations on climate change.

International Institute for Sustainable Development, 161 Portage Avenue East, 6th Floor, Winnipeg, Manitoba, Canada R3B 0Y4. Tel: +1 204 958 7700; Fax: +1 204 958 7710; E-mail: info@iisd.ca

The International Institute for Sustainable Development is concerned with advancing policy recommendations on climate change, as well as international trade and investment, economic policy, and natural resource management, to make development sustainable. Engaged at both the national and international levels in addressing climate change and adaptation, it co-ordinates the Climate Change Knowledge Network and, in collaboration with ENDA-Energy, the Climate Change Capacity Project-Africa.

Environment and Development Action in the Third World (ENDA-TM), 4-5 Rue Kléber, BP 3370, Dakar, Senegal. Tel: +221 (8) 21 60 27 / 22 42 29; Fax: +221 (8) 22 26 95; E-mail: enda@enda.sn; Website in English and French at: http:// www.enda.sn

ENDA-TM is an international NGO based in Dakar, Senegal. It is an association of autonomous entities co-ordinated by an Executive Secretariat, and includes teams and programmes focused on various themes in development and environment. The work carried out by Enda's Energy Programme is centred on the use and development of energy in Africa. The programme uses the principles of research-

action and training to help put into place the United Nations Conventions on Climate Change and Desertification in Africa, and to develop alternative energy technology. Working in partnership, through co-ordinating networks and jointly-led projects, is an important aspect of Enda Energy's work. The team collects information on energy, and implements local-level as well as regional-level projects.

Stakeholder Forum for Our Common Future, 3 Whitehall Court, London SW1A 2EL, UK; Tel: +44 (0)20 7839 1784; Fax +44 (0)20 7930 5893; http://www.stakeholderforum.org

The Stakeholder Forum for Our Common Future, formerly the UNED Forum, was established in 1998 as an international initiative to support international civil society organisations concerned with sustainable development. It includes organisations representing all the major groups recognised by the UN including business, labour, parliamentarians, local government, NGOs, indigenous peoples, women, youth, farmers, and scientists. The United Nations Association of Great Britain and Northern Ireland provides the secretariat for the forum. The forum's work in preparation for Earth Summit 2002 comprises building partnerships and networks, researching and influencing policy, providing and disseminating information, training, and capacity-building.

Women's Environment and Development Organization (WEDO), 355 Lexington Avenue, 3rd Floor, New York, NY 10017-6603, USA. Tel: +1 212 973 0325; Fax: +1 212 973 0335; E-mail: wedo@wedo.org; http://www.wedo.org

WEDO is an international advocacy network concerned with increasing the power of women worldwide as policy-makers in policy-making institutions, forums, and processes at all levels, to achieve economic and social justice, human rights, and a sustainable environment for all. The Sustainable Development Program focuses on integrating gender issues into the global environmental movement by: strengthening international networking between women's and environmental groups; mobilising women's involvement in environmental and sustainable development decision-making; and advocating for gender mainstreaming in international forums like the World Summit on Sustainable Development.

Commission on Sustainable Development (CSD) NGO Women's Caucus
http://www.earthsummit2002.org/wcaucus/csdngo.htm

The CDS Women's Caucus grew out of the 1991 Miami Conference, organised by Women's Environment and Development Organisation (WEDO), and its outcome document, 'Women's Action Agenda for a Healthy Planet'. It is a working group of women and men who are working towards the mainstreaming of gender into sustainable development policies and practical implementation strategies. The women's caucus meets at the CSD Sessions, organises side events, and takes part in other caucuses to ensure gender mainstreaming of NGO work. It operates an open listserve to prepare positions and lobbying strategies, networking globally with interested organisations and individuals.

ENERGIA
http://www.sms.utwente.nl/energia/index.html

Founded in 1995 by an informal group of women involved in energy inputs at the Beijing Conference on Women, ENERGIA is an international network on women and sustainable energy, linking individuals and groups concerned with energy, environment, and women. ENERGIA aims to strengthen the role of women in sustainable energy development through information exchange, training, research,

advocacy, and action. Its quarterly newsletter, ENERGIA NEWS, includes useful resources, and can be accessed through its website.

Women's Environmental Network, P.O. Box 30626, London E1 1TZ, UK. Tel: +44 (0)20 7481 9004; Fax: +44 (0)20 7481 9144; E-mail: info@wen.org.uk
http://www.wen.org.uk

WEN campaigns on issues linking women, the environment, and health. Current campaign issues include health, local foods, nappies, waste, genetic engineering, and sanitary products.

United Nations Development Programme
http://www.undp.org

The Sustainable Energy and Environment Division (SEED) of UNDP contains several papers, reports, and tool kits on gender mainstreaming in natural resource manage-ment. Most materials are available in French, English, and Spanish.

Videos

Questions of Difference – Participatory Rural Appraisal (PRA), Gender and the Environment (1996), Rosanna Horsley for the International Institute for Environment and Development (IIED), 3 Endsleigh St., London WC1H 0DD, UK. Tel: +44 (0)20 7388 2117; Fax: +44 (0)20 7388 2826

This training video draws on experiences from PRA training workshops held in Brazil, Burkina Faso, and Pakistan. It is part of a trainer's pack, which includes a trainer's guide and slide set. It explores the links between gender and the environment through case studies of biodiversity in Brazil, drylands use in Burkina Faso, and mangrove use in Pakistan. Available in English, French, and Portuguese.